Solving Everyday Problems with the Scientific Method

Thinking Like a Scientist

Solving Everyday Problems with the Scientific Method

Thinking Like a Scientist

Don K Mak
Angela T Mak
Anthony B Mak

World Scientific

NEW JERSEY • LONDON • SINGAPORE • BEIJING • SHANGHAI • HONG KONG • TAIPEI • CHENNAI

Published by

World Scientific Publishing Co. Pte. Ltd.
5 Toh Tuck Link, Singapore 596224
USA office: 27 Warren Street, Suite 401-402, Hackensack, NJ 07601
UK office: 57 Shelton Street, Covent Garden, London WC2H 9HE

Library of Congress Cataloging-in-Publication Data
Mak, Don K.
Solving everyday problems with the scientific method : thinking like a scientist / by Don K. Mak, Angela T. Mak, Anthony B Mak.
p. cm.
Includes bibliographical references and index.
ISBN 978-981-283-509-3 (hardcover)
ISBN 978-981-4304-04-7 (paperback)
1. Science--Methodology. 2. Problem solving--Methodology. I. Mak, Angela T. II. Mak, Anthony B. III. Title.

Q175.M258 2009
501--dc22

2008052087

British Library Cataloguing-in-Publication Data
A catalogue record for this book is available from the British Library.

First published 2009
Reprinted 2010

Printed in Singapore.

In memory of

My uncle, Mr. Mak Chung Lun, a kind-hearted gentleman, who was separated from his wife after only a few years of marriage[+].

DKM

[+] His observant mother, Ms Chow Chu, repeatedly cautioned him not to marry his wife. Even on the night before his wedding, she pleaded with him, "It is still not too late. I cannot stay with you for the rest of your life. You are the one who will be living with your spouse."

Her scientific premonition proved to be true.

Claimers and Disclaimers

The events given as examples in this book have actually occurred. However, the names of the people, places, as well as some of the minor details have been changed to protect privacy.

Solutions of some medical problems mentioned in this book may not work for everyone. Patients should observe, hypothesize, and experiment under the supervision of medical doctors.

Preface

Bunny was born a happy baby. She spent her whole day playing, eating, and sleeping. There was nothing to worry about. Life was great.

As time went on, she grew a bit bigger, and she was more aware of her surroundings. And she had to take responsibility to take care of herself. Events did not always happen the way that she wanted. Problems occurred that she did not know how to handle. Life became miserable.

One day, Bunny met Mr. Rabbit. Mr. Rabbit is a wise sage. He listened to Bunny's difficulties. He understood what her obstacle was. So he taught her The Scientific Method. Not only would The Scientific Method help her solve problems in situations that she is familiar with, it would also improve her thinking skills in environments that she is not accustomed to.

Bunny learnt The Scientific Method, and she practices it every day if and when she gets a chance. She is able to solve more problems than she ever could. And she lives happier ever after.

Contents

Chapter 1

Prelude

The father put down the newspaper. It had been raining for the last two hours. The rain finally stopped, and the sky looked clear. After all this raining, the negative ions in the atmosphere would have increased, and the air would feel fresh. The father suggested the family of four should go for a stroll. There was a park just about fifteen minutes walk from their house.

The mother got their three-year-old son and five-year-old daughter dressed. They arrived at the park, and rambled along the path leading to the playground. Not exactly watching where she was going, the daughter stepped one foot into a puddle of water. Both her sock and the shoe got wet. She refused to walk any further. Even with some persuasion, she declined to walk. The father pondered what action should he be taking. Shall I carry her all the way back home? I may get a backache or a hernia. Maybe I shall run home and get the car. Or, maybe, I should force her to continue walking. Which path should I choose to solve this problem?

Take a minute to think what you would suggest. And before we find out how the father is going to cope with the situation, let's see what exactly the scientific method is all about.

Chapter 2

The Scientific Method

In the history of philosophical ideation, scientific discoveries, and engineering inventions, it has almost never happened that a single person (or a single group of people) has come up with an idea or a similar idea that no one has ever dreamed of earlier, or at the same time. This person may not be aware of the previous findings, nor someone else in another part of the world has comparable ideas, and thus — his idea may be very original, as far as he is concerned. However, history tells us that it is highly unlikely that no one has already come up with some related concepts.

The ideation and development of the scientific method is no exception. No single person, a group of people, or a certain civilization can claim the credit for inventing the scientific method. The method is slowly evolved through centuries. It may have started with cavemen using their stone tools. There are, however, some significant milestones along the way.

2.1 Edwin Smith papyrus

The origin of the scientific method can be traced back to approximately 2600 BC. Ancient surgical methods were documented in the Edwin Smith papyrus, a manuscript bought by an Egyptologist Edwin Smith in 1862 in Egypt. Papyrus is an aquatic plant native to the Nile valley in Egypt. The spongelike central cylinders of the stems of the plant can be laid together, soaked, pressed, and dried to form a scroll, which was used by ancient Egyptians to write on. Imhotep (circa 2600 BC), the founder of Egyptian medicine, is credited as the original author of the Edwin Smith papyrus, which is considered to be the world's earliest known medical document. The document compiles a list of forty-eight battlefield injuries, and the prudent surgical treatments that the victims had received. It describes the brain, heart, liver, spleen, kidneys, and bladder. It also depicts surgical stitching and different kinds of dressings. The papyrus contains the essential elements of the scientific method: examination, diagnosis, treatment, and prognosis.

2.2 Greek philosophy (4th century BC)

Another significant contribution to the scientific method occurred in the fourth century BC in Ancient Greece. One of the key figures was the Greek philosopher, Aristotle (384–322 BC). Aristotle was born in Stagira, which was near Macedonia. His father was the family physician of King Amyntas of Macedonia. From his father, Aristotle received training and knowledge that would encourage him toward the investigation of natural phenomena.

When he was seventeen, he was sent to study at Plato's Academy in Athens, which was the largest city in Greece. At that time, the Academy was considered as the centre of the intellectual world. He stayed there for twenty years, until the death of Plato (427–347 BC). Nevertheless, Aristotle disagreed with Plato on several basic philosophical issues. While Plato believed that knowledge came from conversation and methodical questioning, an idea that originated from his teacher Socrates (469–399 BC), Aristotle believed that knowledge came from one's sensory experiences. Plato theorized that, through intellectual reasoning, the laws of the universe could be discovered. However, Aristotle attempted to reconcile abstract thought with observation. While both Plato and Aristotle supported deductive reasoning, only Aristotle championed inductive reasoning

Deductive reasoning is a logical procedure where a conclusion is drawn from accepted premises or axioms. A logic system, now sometimes called the Aristotelian logic, has been developed by Aristotle. One famous example is: from the two statements "Human beings are mortal" and "Greeks are human beings", we can come to the conclusion that "Greeks are mortal".

Inductive reasoning starts with observations, from which a general principle is derived. For example, if all swans that we have observed are white, then we can come up with the generalization that "All swans are white". If someone tells us that he just saw a swan running along the street, we can deduce (i.e., using deductive reasoning) that the swan must be white in color. However, we need to be careful in our observations before we come to any general principle. For instance, if we ever see a black swan in the future, we would need to discard our general principle.

Aristotle had a wide interest and wanted to know just about everything in nature. If there were something that he did not understand, he would attempt to discover the answer by making observations, collecting data, and thinking it through. However, he did make some occasional mistakes. For example, he said that women had fewer teeth than men had. He also wrote that a king bee, not a queen bee, ruled the hive. While he stressed on observation, he did not attempt to prove his theories by performing experiments. For instance, he claimed that heavy objects fell faster than light objects. This proposition was later refuted by the Greek philosopher, John Philoponus (~490—~570 AD). Centuries later, Galileo (1564–1642 AD) established experimentally that heavy objects fell at practically the same rate as light objects. Aristotle also failed to see the application of mathematics to physics. He thought that physics dealt with changing objects while mathematics dealt with unchanging objects. That conclusion obviously would have affected his perception of nature.

Aristotle had written about many subjects, viz., ethics, politics, meteorology, physics, mathematics, metaphysics, embryology, anatomy, physiology, etc. His work exerted a lot of influence in later generations. For example, his books on Physics were served as the basis of natural philosophy (now known as natural science) for two thousand years, up to the era of Galileo in the sixteenth century.

It had been asserted that Aristotle's writings actually held back the advancement of science, as he was so respected that he was often not challenged. However, he did explicitly teach his students to find out what had previously been done on a certain subject, and identify any reasons to doubt the beliefs and come up with theories of their own. Nevertheless, his fault was that he did not perform any experiment to validate his theories.

2.3 Islamic philosophy (8th century AD–15th century AD)

Muslim scientists played a significant role in the development of the scientific method in the modern form. They placed more emphasis on experiments than the Greeks. Guided by Islamic philosophy and religion, the Muslim's empirical studies of nature were based on systematic observation and experimentation. Muslim scholars benefited from the use of a single language, Arabic, from the newly created Arabic community in the 8th century. They also got access to Greek and Roman texts, as well as Indian sources of science and technology.

The prominent Arab scientist, Ibn Al-Haitham (known in the West as Alhazen) (965 AD–1040 AD), applied the scientific method for his optics experiments. He examined the passage of light through various media, and devised the laws of refraction. He also performed experiment on the dispersion of lights into its component colors. His book, the Book of Optics, was translated into Latin, and has exerted a great influence upon Western Science.

Another distinguished scientist, Al-Biruni (973 AD–1048 AD), contributed immensely to the fields of philosophy, mathematics, science and medicine. He measured the radius of the earth, and discussed the theory of the earth rotating about its own axis. Furthermore, he made reasonably precise calculation of the specific gravity of eighteen precious stones and metals.

Similar scientific studies were carried out by Muslim scientists in a much wider scale than had been performed in previous civilizations. Science was then, an important discipline in the Islamic culture.

2.4 European Science (12th century AD–16th century AD)

With the fall of the Western Roman Empire in 476 AD, a large portion of knowledge of the past was lost in most of Europe. Only a few copies of ancient Greek texts remained as the basis for philosophical and scientific learning.

In the late 11th and 12th centuries, universities were first established in Italy, France, and England for the study of arts, law, medicine, and theology. That initiated the revival of art, literature, and learning in Europe. Through communication with the Islamic world, Europeans were able to get access to the works of Ancient Greek and Romans, as well as the works of Islamic philosophers. Furthermore, Europeans began to travel east, leading to the increased influence of Indian and even Chinese science and technology on the European scene.

By the beginning of the 13th century, distinguished academics such as Robert Grosseteste and Roger Bacon began to extend the ideas of natural philosophy described in earlier texts, which had been translated into Latin.

Robert Grosseteste (1175–1253), an English philosopher, had written works on astronomy, optics and tidal movements. He had also written a few commentaries on Aristotle's work. He thoroughly comprehended Aristotle's idea of the dual path of scientific reasoning (induction and deduction), that discussed the generalizations from particulars to a general premise, and then using the general premise to forecast other particulars. However, unlike Aristotle, Grosseteste accentuated the role of experimentation in verifying scientific facts. He also emphasized the importance of mathematics in formulating the laws of natural science.

Roger Bacon (1214–1294) was a Catholic priest and an English philosopher who thought mathematics formed the base of science. He was quite familiar with the philosophical and scientific works in the Arab world. Like Grosseteste, he placed considerable emphasis on acquiring knowledge through deliberate experimental arrangements, rather than relying on sayings from authorities. An experiment had to be set up as a test under controlled conditions to examine the validity of a hypothesis. If the conditions were controlled in precisely the same way in a repeated experiment, the same results would occur. All theories needed to be tested through observation of nature, rather than depending solely on reasoning and thinking. He was considered in the West as one of the earliest advocates of the scientific method. He has written topics in mathematics, optics, alchemy, and celestial bodies.

In the 14th century, an English logician, William of Ockham (1285–1349) introduced the principle of parsimony, which is now known as the Ockham's Razor. The principle states that an explanation or a theory should be as simple as possible and contains just enough terms to explain the facts. The term "razor" is used to mean that unnecessary assumptions need to be shaved away to obtain the simplest explanation. The Razor is sometimes stated as "entities are not to be multiplied beyond necessity". It parallels what Einstein wrote in the 20th century, "Theories should be as simple as possible, but not simpler".

In the year 1347, a devastating pandemic, the Black Death, struck Europe, and killed 1/3 to 2/3 of the population. Simultaneous epidemics also occurred across large portions of Asia (especially in India and China) and the Middle East. The same disease was thought to have returned to Europe for several generations until the 17th century. This drastically curtailed the flourishing philosophical and scientific development in Europe. However, the introduction of printing from China during that period had a great impact on the European society. Printing of books changed the way information was transferred in Europe, where before, only handwritten manuscripts were produced. It also facilitated the communication of scientists about their discoveries, thus bringing on the Scientific Revolution.

2.5 Scientific Revolution (1543 AD–18th century AD)

The Scientific Revolution was based upon the learning of the universities in Europe. It can be dated as having begun in 1543, the year when Nicolaus Copernicus published *On the Revolutions of the Heavenly Spheres*. The book contested the universe proposed by the Greek astronomer Ptolemy (90–168 AD), who believed that the earth was the centre point of the revolution of the heaven.

Ptolemy and some other astronomers believed that the planets moved in concentric circles around the earth. However, sometimes the planets were observed to move backwards in the circles. This was described as a retrograde motion. To interpret this behavior, planets were depicted to be moving, not on the concentric circles, but on circles with centres that were moving on the concentric circles. These smaller circles were called epicycles. While the planets moved in a uniform circular motion on the epicycles, the centres of the epicycles moved in uniform circular motion around the earth. This could explain the retrograde motion.

In order to explain the detailed motions of the planets, sometimes epicycles were themselves placed on epicycles. In Ptolemy's universe, about 80 epicycles were used to explain the motions of the sun, the moon, and the five planets known in his time. This description fully accounted for the motions of these heavenly bodies. Nevertheless, when King Alfonso of Castile and Leon was introduced to Ptolemy's epicycles in the thirteenth century, he was so baffled that it had been said that he commented, "If God had made the universe as such, he should have consulted me first". As a matter of fact, even Ptolemy himself did not like this clumsy system. He argued that his mathematical model was only used to explain and predict the motions of the universe. It was not a physical description of the universe. He stated that there could be other equivalent mathematical model that could yield the same observed motions.

Nicolaus Copernicus (1473–1543) was the first influential astronomer to question Ptolemy's theory that the earth was the centre of the universe. He proposed that the sun was actually the heavenly object where the earth and other planets revolved around in circular orbits. While his system was simpler, he still needed epicycles to explain the retrograde motions of the planets.

It was Johannes Kepler (1571–1630) that pointed out that the planets actually revolved around the sun in elliptical orbits, with the sun being at one focus. An ellipse is a flattened circle. The sum of the distance from any point on an ellipse to two fixed points is a constant. The two fixed points are called foci (singular: focus). In the Keplerian universe, epicycles were eliminated. This was a significant improvement over the Copernican universe. His improved model explained all planetary motions, including the retrograde motions of the planets.

Kepler made use of the extensive data collected by Tycho Brache (1546–1601), with whom he worked as an assistant. Brache had made very accurate observation. However, he hypothesized that the earth was the centre of the universe. It was Kepler who had the insight to propose the correct theory. While precise observations are important, one needs testable theories to account for the observation in a logical and mathematical manner. The interplay between observation and theoretical modeling depicts the development of modern science.

Now, there is one more problem that needs to be solved. Since the earth is revolving around the sun, it must be moving very fast. If we are to jump up, we should be landing on the earth away from the location where we jump. However, that is not the case. The explanation was provided by Galileo (1564–1642), who discovered the law of inertia during the first decade of the seventeenth century. The law states that if an object is moving at a constant speed in a certain direction, it will continue to move at that speed in that direction, as long as no force in the motion's direction is acting on it.

Galileo believed in the sun-centred system rather than Ptolemy's earth-centred system. In addition, he advocated that the former was a physical model that represented reality, and not necessarily a mathematical model (with false physical axioms) as suggested by Copernicus. That is, the sun was actually the centre of the universe. This proposition did not sit well with the Roman Catholic Church, who considered his argument contradictory to the church doctrine. The church demanded him to recant his ideas, and he was put under house arrest. Nevertheless, his idea that the physical model should be consistent with the mathematical description of a phenomenon would eventually form the basis of scientific development in the modern world. A physical model, not only should it predict the behavior of the world, should also give us insight into its nature.

The physical model of the universe was provided by Isaac Newton (1642–1727), who introduced the Law of Universal Gravitation. The earth and the planets revolved around the sun through gravitational attraction between them and the sun. Newton was the greatest scientist of his era. He conducted many experiments and was responsible for the immense advancement of our understanding of mechanics and optics. He wrote Principia Mathematica in 1687 and Opticks in 1704.

Other disciplines were also flourishing during the Scientific Revolution. In 1543, Andreas Vesalius (1514–1564), a Belgian physician, published *On the Fabrics of the Human Body*. The book relied on observation taken directly from human dissection. This was in mark contrast to the writings of the Greek physician, Galen (129–200), who could only dissect on animals, mostly apes. As humans differed in anatomy from animals, Vesalius's book was the most accurate and comprehensive anatomical text at his time.

In 1665, Robert Hooke (1635–1703) published Micrographia, which is the first book describing observation made through a microscope. Hooke had devised a compound microscope (i.e., a microscope using more than one lens), and used it to observe organisms like insects, sponges, and cork. He was the first person to use the word "cell" to describe the microstructure in cork.

Inspired by Micrographia, Antony van Leeuwenhock (1632–1723), a Dutch tradesman, learned how to grind lenses and build simple microscopes (with a single lens) with magnification of over 200 times. His microscopes were more powerful than Hooke's compound microscope, which could magnify only to about thirty times. With his microscopes, he was the first to see bacteria in a drop of water, and blood corpuscles in capillaries. He studied a broad range of living and non-living microscopic phenomena, and reported his findings to the Royal Society of England, which is an independent scientific academy dedicated to promoting excellence in science.

At the end of the Scientific Revolution, knowledge was no longer dictated by authorities, but accumulated painstakingly by experimental research. All these have been made possible through the introduction of philosophical ideation in humanism and empiricism for the past centuries.

2.6 Humanism and Empiricism

Humanism emphasizes reason and scientific inquiry in the natural world. It is based on the idea that human intellect is reliable to acquire knowledge and human experience can be trusted. The idea started as early as the 6th century BC. The pre-Socratic Greek philosopher, Thales of Miletus (about 624 BC–about 546 BC), proposed theories to explain many of the events of nature without reference to the supernatural. He was credited with the quotation "Know thyself". Before Thales, the Greek explained phenomena like lightning and earthquake as actions from the Gods. Thales explained earthquakes as the earth being rocked by the water that it floated upon. Even though the explanation was not correct, he did attempt to attribute these natural phenomena as caused by nature. However, humanism was quite often challenged. For example, even in the beginning of the seventeenth century, Galileo was put on trial for his proposition of the sun-centred universe, and he had to choose between believing in his observation or the teaching of the Church.

Galileo practiced empiricism, and performed experiments. Empiricism is the doctrine that emphasizes the aspect of knowledge that is derived from one's sense experience, especially through experimentation. In the Aristotelian scientific era, conclusions about nature were drawn from observations of natural phenomena. Experiments were seldom performed, if at all. Aristotle's "laws of the universe" were somewhat qualitative and faulty. His incorrect theory that heavy objects fell faster than light objects could have been discarded if an experiment had been performed.

The English philosopher, Francis Bacon (1561–1626), criticized Aristotle's method of induction as coming to conclusion of a general proposition too quickly and only from a few observations. He introduced his "true and perfect" induction method, which consisted of a ladder of axioms, with the most general and comprehensive axiom at the top and the most narrow and specialized axioms at the bottom. Each step had to be thoroughly tested by observation and experiment before the next step was taken. The Baconian method involved the careful collection and interpretation of data from detailed and methodical experimentation. While the method would lead to a very systematic accumulation of information, it was often criticized for its underestimation of hypothesis.

Hypothesizing requires a leap from observed particulars to abstract generalizations, which is set forth to explain the phenomena. Imagination is necessary to attain a breakthrough in scientific discoveries. For example, while Tycho Brache behaved like a Baconian and painstakingly recorded detailed astronomical data in tables, it was the imaginative thinking of Kepler to figure out that the planets actually moved around the sun in elliptical orbits.

All these accumulated experience of philosophers and scientists eventually form the basis of the scientific method, which is the tool of modern scientific investigation.

2.7 The Scientific Method

Now, what exactly is the scientific method? It can actually be described in different versions.

A comprehensive version can be depicted as such: Observation, Recognition, Definition, Hypothesis, Prediction, and Experiment.

Observation is the noticing or perceiving of some aspect of the universe. Then, one needs to recognize that a problem-situation is significant enough to require attention. The circumstance is then defined or modeled. A tentative description or hypothesis is then formulated to explain the phenomenon, and to predict the existence of other phenomena. The prediction is then tested by an experiment.

The hypothesis may be accepted or modified or rejected in light of new observations. A hypothesis has to be capable of being tested by an experiment, i.e., it has to be falsifiable. This differentiates it from a belief or a faith. Thus, the statement "This is destiny" is not falsifiable, as no experiment can be designed to prove whether it is true or not. The strength of a hypothesis is in its predictive power — where we can get more out than what we have put in. The validity of a hypothesis has to be tested under controlled conditions. In its simplest form, a controlled experiment is performed when one variable (the independent variable) is changed, thus causing another variable (the dependent variable) to change at the same time. All other variables will be kept as constants. The result of the experiment has to be reproducible by others under the same experimental description and procedure.

This comprehensive version of the Scientific Method can be abbreviated to: Observation, Hypothesis, and Experiment. This simple version seems to suffice for accomplishing quite a number of scientific works, and also for coping with everyday problems.

2.8 Application of the Scientific Method to Everyday Problem

Everyday problems share the same commonalities as scientific problems. They are situations that require solutions; they hold difficulties that need to be resolved. As such, everyday problems would benefit by employing the scientific method. We will study how the scientific method can be used in daily life.

Example

Let us take a look at the wet foot problem discussed in Chapter 1. The father noticed that his daughter stepped only the left foot into the puddle of water. As a result, only the sock and the shoe of the left foot got wet. After drying the left foot first, he took off her right sock and put it onto her left foot. He then put both shoes back on, leaving the right foot without a sock. The little girl felt comfortable, and did not complain. The whole family continued walking to the playground. The children spent half an hour playing there, and the family then walked back home after.

Here we note that the father *observed* where the problem was. He *hypothesized* what could be the solution. He *tested* it out, and found that the idea worked.

Chapter 3

Observation

Observation is the first step of the Scientific Method. However, it can infiltrate the whole scientific process — from the initial perception of a phenomenon, to proposing a solution, and right down to experimentation, where observation of the results is significant.

In daily life, observation is equally important. We should anticipate problems before they arise, and seek solutions after the obstacles have occurred. In addition, we need to be always on the look out for opportunities, and search for various avenues to obtain betterments. Thus again, we can see that from the recognition of a problem, to the finding of its solution, observation is essential.

Tom had been away on a business trip for two weeks. When he came home, and entered through the front door, he realized that the screws on the doorknob were loose. He thought to himself that this would make it easy for someone to break into the house. He quickly got a screwdriver to tighten them. His wife had come into the house through that same door everyday for the last two weeks, but she had not noticed that the screws were loose. She was not a very observant person and did not recognize that there was a problem to begin with.

Recognition of a problem is actually a prelude to solving the problem. We need to realize that a problem situation has occurred. This may sound easy, but some problems are hidden and may not be easily spotted. This is why we should train ourselves to be alert to our surroundings.

Observation does not necessarily mean that we have to see it with our own eyes. We have five senses. Sight is only one of them. The others are hearing, taste, touch, and smell. Can we hear a noise coming out from our car engine? Does the soup taste funny? Should we buy a bath towel that is rough to the touch of our fingers? Do we smell something burning in the oven?

Once a problem is recognized, observation is required to find a solution, from whatever information we can gather from our five senses. Information can also be obtained from various sources — reading books, our own past experience, talking to people, searching the Internet, etc. Hopefully, the knowledge gathered can provide a hint to solving the problem. We will now take a look at a few examples and see how observation has solved some of the real life problems.

Example 1 Indigestion

While Raymond was doing his undergraduate at University of Syracuse, his eldest sister, Diana, was doing her Master Degree in Social Work at University of Michigan in Ann Arbor. One weekend, Diana came to visit her brother, Raymond.

As it happened, for the two weeks before she came, Raymond had been burping a few times a day. It was annoying, but it did not really bother him, nor did he pay much attention to it.

Three days after Diana stayed with her brother, she suggested to him that he should cut down on oranges. She heard him burp, and she saw him eating oranges. And she saw the correlation. It suddenly dawned on Raymond that his sister might be right. For the last two weeks, Raymond had been eating two oranges a day instead of one a day as he did before. He read it somewhere that one orange contained approximately 50 mg of Vitamin C, and as he thought his daily dose of Vitamin C should be 100 mg, he started eating two oranges a day. Orange juice was acidic due to its high citrus acid content, and his stomach might not have tolerated it. He never realized the problem, and never saw the correlation between oranges and his burping. It was fortunate that his sister was there, and pointed it out. He cut back to eating one orange per day. His burping disappeared a couple of days later.

Example 2 Restaurant Christmas menu

It was about one week before Christmas. The father took the family of four out for a holiday dinner at a restaurant. As it was Christmas time, the restaurant had a special one-page menu catered for the festival. The father looked at the menu, and would like to test his teenager children to see how observant they were. He had been training them to solve problems since they were kids. He asked them to take a look at the menu and see whether they noticed anything particularly interesting in it.

The daughter looked at the menu and noticed that there were some special dishes that were different from the dishes in the regular menu. She wanted to order the Thai curry chicken from the special menu. The son wanted to order the plum-glazed pork ribs.

However, those special dishes were not what the father had in mind. He then pointed out to his children that at the bottom of the menu was a note, which said that the purchaser of a $50 gift certificate of the restaurant would receive two complimentary glasses, which were a promotional offer from a beer company.

They ordered their food, and had a pleasant meal. After they finished their dinner, the father asked for the bill. The bill, including tips, came out to be about $100. The father asked the waitress whether they could take a look at the glasses, which turned out to look quite pleasing. He then purchased two $50 gift certificates, paid the bill with those certificates, and took four glasses home.

Example 3 Multi-vitamins

Richard looked at the drugstore flyer that had been delivered to his house. The drugstore was having a sale on a certain brand of multi-vitamins. As he would need some multi-vitamins, he went to the drugstore to buy some.

The multi-vitamin was contained in a plastic bottle that was placed inside a paper box. The expiry date was imprinted on the box that had a white background, making it very difficult to see what the expiry date was. Nevertheless, Richard discerned that one bottle had expired a few months ago and the rest would expire in the next month. He notified the drugstore clerk, who instantly removed the one that had expired, leaving the rest on the shelf.

Obviously, Richard did not buy any of those multivitamins. As he was leaving the store, he was wondering whether a not-so-observant consumer would be purchasing those soon-to-be-expired multivitamins.

In order to solve a problem, we need information. You may have heard people say that one can start off with a blank piece of paper, so that one may not be biased with preconceived ideas. But that is a misconception. No one can create something from nothing.

The information that we are going to make use of can be external or internal. External means that we need to find it. Internal means that we already have it stored in our brains, but we need to extract them to cope with the problem at hand. More often than not, we need to use a combination of external and internal information. Let us take a look at external information first.

3.1 External information

We need to make observation of our surroundings in order to find the data that we require. Keen perception is most important. Not paying attention can be costly.

3.1.1 *Missed information*

Occasionally, there is information that we should be aware of, but somehow escape our attention, as the following example will show.

Example 4 Car accident

The Jones live in Cornwall, Canada. One day, the teenager daughter backed the car out of the driveway, and hit their neighbour's car that was parked across the street. The neighbour called the police to report the incident. The police came and after inspecting the damage, charged her with 6 demerit points penalty for careless driving. The father also later had to pay the neighbour $600 to cover the damage.

The daughter admitted that she did not look before backing out the car. The father told her that she should have surveyed the surrounding, including the area that she backed the car up to, even before she got into the car. In addition, she should be monitoring the space behind the car, as she was backing out, because information changes all the time. There might be another car coming down the road, or a child running along the street. Not observing certain information and not realizing the time dependence of information can be very costly, as shown in this particular case.

Missed information is unfavorable for judgment and decision making. Unfortunately, there are situations that can be worse in principle and in practice. We can be misled by incorrect information that is provided by others. We will describe this in the next section.

3.1.2 *Misinformation*

Sometimes, we are given false or misleading information, unintentionally or intentionally. If we are doubtful about the information, we should check it out in other avenues. However, if we are not aware that the information is incorrect to begin with, we will accept it as it is until we find out otherwise later.

Let us take a look at some examples.

Example 5 Vinyl flooring installation

In the summer of 2005, Lucy wanted to replace her kitchen floor with a new vinyl flooring as well as putting in a new baseboard. Baseboard is a piece of wooden board, usually about 10 cm high, used to cover the lowest part of an interior wall, so as to conceal the joint between the wall and the floor.

She hired a home renovation company to install the flooring as well as the baseboard. The company sent someone over to take the measurement of her kitchen, and gave her a quotation of the material and labour of the length of baseboard and the area of the vinyl flooring required. The whole job would cost about $1,000.

As Lucy wanted to purchase and pre-paint the baseboard herself before it was installed by the company, she measured the dimension of her kitchen herself, so as to double check how much baseboard she needed to buy. Somewhat to her surprise, the length of the baseboard came out to be only 69 feet, which was 10.1% less than the 76 feet in the company's quotation. She called and mentioned the discrepancy to the company, which then suggested that she asked the installers to re-measure the kitchen when they came to install the flooring. The installers confirmed that Lucy was correct in her measurement, and she was eventually able to get a refund of about $25 from the company.

A year later, a friend of hers, Nancy, wanted to hire the same company to install hardwood flooring in her living room. Lucy then told Nancy about her experience with the company. Nancy, being a more mathematically oriented person than Lucy, quickly pointed out that if the linear measurement was inflated by 10.1% (= 0.101), the area measurement would have been inflated by

$$(1 + 0.101) \times (1 + 0.101) - 1 \approx 0.21 = 21\%.$$

As the vinyl flooring material and installation labour cost about $700, that means that the company had over-charged Lucy by $0.21/(1 + 0.21) \times \$700 \approx \$121$. Lucy checked the record of her measurement of the kitchen floor, and agreed that Nancy was indeed correct. However, she decided not to file a complaint as the job was done a year ago.

This example just shows that misinformation can cost the consumer extra money. It also shows that the knowledge of some mathematics would definitely be helpful. We would elaborate further on this in the chapter on Mathematics.

Example 6 Guided tour

A couple joined a guided group tour to go to Thailand. The tour included a one-hour traditional Thai massage in a massage parlor. As some other members in the group opted to pay for one more hour of massage, the couple had one hour to spare, and they wandered into the dried food store next door.

The store was selling pork and beef jerky. Samples were displayed on the table for customers to taste. A jerky is a piece of meat that has been cut into strips, marinated, and dried under low heat (usually under 70° C). It is considered a delicacy in Thailand. The couple tried some samples, and thought they tasted quite good. So they bought 1 Kg of beef jerky. While they were wondering whether they should buy more and gave some to their relatives as a gift, they ran into the tourist guide who told them that they would be going to Store X the next day, and Store X's jerky tasted much better. So, the couple decided to wait.

The next day, the tourist guide took the whole group to shop at Store X. The couple sampled the jerky there. Not only did the jerky taste not as good, the price was 25% more than that of the store they went to the day before. They bought 2 Kg of jerky anyway, because they wanted their relatives back home to try some of the delicacies in Thailand.

They later found out that the reason the guide recommended Store X was because she got a kickback commission from that store.

Thus, the couple learned their lesson, and became smarter in the future. From then on, they would scrutinize where the information is coming from, and whether the person giving out the information has any conflicting interest. This experience made them more alert to what they were being told when the whole family of four traveled to Europe a few years later.

Example 7 Hotel breakfast

In the summer of 2007, a family of four planned to go to tour Europe for about four weeks. They asked their travel agent to rent a car, and book the hotels for them. In total, nine hotels were booked, with two rooms in each hotel. The travel agent asked them whether they would like to book breakfasts in the hotels as well. She was told that they did not particularly care for breakfast. However, they would eat breakfast at the hotel if the breakfast was complimentary. After all the booking was done, the travel agent gave them a printout of the hotels, the hotel rates, their addresses and amenities. In the printout was also listed which hotels served complimentary breakfast.

The family arrived in Berlin just before midnight, and everybody was tired. They checked in at the hotel, and were told by the clerk at the front desk that their hotel rate included breakfast, and the hotel started serving breakfast at seven in the morning.

When the father got to his room, he checked the printout and found that breakfast was not complimentary in that hotel. The hotel rate for each room was listed at 100 Euros. He then called the front desk and asked how much was the hotel rate that they were being charged per room, and was told that the rate was 130 Euros, which included 15 Euros breakfast per person. He therefore informed the front desk clerk to cancel breakfast. It appeared that someone had intentionally or unintentionally altered the hotel rate to include breakfast. As there were four members in the family, and they were staying in Berlin for four nights, they would have wound up paying 240 Euros (approximately US $350) more to the hotel.

Example 8 Overweight products in supermarket

While Robert was shopping in the supermarket, he saw that chicken was on sale for $0.99/lb ($2.18/Kg). So he bought four. He took them back home and put them in the freezer. A week later, he took one of them out to defrost for supper. He looked at the label, which said that the chicken weighted over 7 lb. As he did not feel that the chicken was that heavy, he weighted it in the digital scale in his bathroom. The display of the scale showed approximately 5 lb, i.e., about 2 lb less than what the label said. He checked the other three chickens. Each one of them weighted in the range of 1 lb to 2 lb less of what its label said. Robert wondered how that could have happened.

In the next half year, he became more attentive to the weights printed on the labels of meat products. He had found several incidents in different supermarkets where the products were overweight, as he could check the weights of the products by putting them on the spring scales set up for measuring produces in the supermarket. He hypothesized that the packaging staff might have thrown the product on their digital scale, and hit the print label button before the scale was settled. That seemed to agree with Newton's Second Law, which stated that the force was equal to the rate of change of momentum of a body, which in this case was the packaged meat.

Once, he saw chicken legs selling at $2.18/Kg in a supermarket. They were all wrapped up in packages of about the same size. While most of them were priced at around $5.50 and weighted approximately 2.5 Kg, he noticed one of them was priced at $7.14 and weighted 3.278 Kg according to the label. He noted that the price/Kg of that package was the same as the others. As he was quite doubtful of whether that was the correct weight, he took the package to the spring scale and found that it actually weighed approximately 2 Kg, instead of 3.278 Kg as labeled. In order to double confirm the weight, he asked the store clerk whether she could take it back to the packaging department to have it re-weighted. The clerk was puzzled, but eventually agreed to do so. Robert saw the packaging clerk gently put the chicken leg package on the scale, and then pressed for the label. The new label read 1.958 Kg and the price was changed to $4.27. Robert thought that his hypothesis regarding Newton's Second Law might be correct. However, it would take more testing to confirm his conjecture.

While some information can be easily found out by us without much difficulty, others are hidden and need to be extracted. In the next section, we will see how we should be aware of any hidden information.

3.1.3 *Hidden information*

In daily life, some information is not so obvious. A well-known, even though fictional example is "the dog did not bark" incident in one of Sherlock Holmes's short stories.

A famous racehorse disappeared, and the trainer was murdered. Both the Scotland Yard detective and Sherlock Holmes had inspected the crime scene. When the detective asked Holmes whether there was any particular detail that Holmes would like to draw his attention, Holmes replied that it was the curious behavior of the dog that night. The detective told Holmes that the dog did nothing that night. Holmes remarked that was exactly it. The fact that the dog did not bark would imply that the intruder was not a stranger.

At first sight, it seems as if the dog did not provide any information at all, but the fact that it did not provide any information is the hidden information that one should be looking out for.

Let's take a look at some real life examples where hidden information is important.

Example 9 Swollen feet

Ron was born in Hong Kong. After finishing high school, he left by himself to go to university in the United States, and eventually settled there. Now that his mother was in her late eighties, he tried to go back to Hong Kong once a year to see his mother, and stayed there for about three weeks. Her mother's birthday was in November. So, Ron usually tried to go back in early November. Brothers and sisters would have a birthday party for their mom. For her age, his mother was reasonably healthy. She exercised often, and knew how to take care of herself.

In November two years ago, Ron flew back to Hong Kong. The plane arrived late at night. When he got to his mom's apartment, he talked briefly to his mom, and then went to bed. He was waken up in the morning by a phone call. It was his aunt. She told him, "You need to save your mom. She told me that she wanted to die". Ron was startled and asked her why. She explained that his mother had skin rashes all over her body and her feet were quite swollen. His mother told her that she did not want to live.

Ron understood that some chronic ailments could drag on for a long time. Even though they were not fatal, they could be so painful and irritating that the patients could lose their will to live. However, Ron knew nothing about medicine. As a matter of fact, he did not even take Biology in his first year of university in Science. Biology was just not his cup of tea. And in any case, he had always believed that his mom had been very well taken care of. His brother-in-law, Prof. Leung, was a medical doctor and a professor at the Chinese University of Hong Kong. He would have had the best connection in town, and would have recommended a good dermatologist to take care of her. And indeed he did. The dermatologist gave her medication and ointment, and asked her to put baby oil in warm water in the bathtub, and soak the whole body in there for half an hour each day. She had been following the doctor's order for the last few months. Unfortunately, that did not seem to help too much.

For the next few days, Ron watched helplessly as her mom put ointment on her feet, which had swollen to about 25% more than their normal size. When she combed her hair, a lot of hair was falling off. She wailed about her hair loss. Even at her age, she still wanted to look pretty. Ron could do nothing about her chronic illness, as he did not know what was happening.

While Ron stayed with his mom, quite often he and his mom had supper at home. His mom's maid made excellent steamed fish. She cooked it just right, and that was much better than the steamed fish cooked in restaurants, where they usually overcooked the fish. One day, while he was having supper with his mom, he saw her scraping the fish skin off the fish. He thought that was strange, but did not make any comment.

A couple of days later, his mom's skin rashes got very itchy to the point that they were unbearable. She exclaimed that she would rather die. Ron then asked, "Mom, when did you start having skin rashes?" His mom told him it started about nine months ago. Ron asked whether anything in particular happened at that time. She told him she had a medical checkup with a general practitioner. As her cholesterol level was a bit on the high side, the doctor told her not to eat the skins of any animals, including fish skin. Ron suddenly realized what was happening. He then said, "Mom, start eating fish skin from now on. I can guarantee that you should get much better in a month and a half. Your diet has been lacking very much in fat." Ron knew there was a risk that his mom's cholesterol level would get higher, but the benefit of eating fish skin would much outweigh the risk. Judging from risk-benefit analysis, he believed that his mom should consume some fish skin.

By chance, her mother was going to see the dermatologist the next day. She double-checked with him whether she should be eating any fish skin. He told her she could. He also told her that while she should be cutting down eating fatty food, she should not completely cut off eating all fats as she had been doing. So, she started eating fish skin from then on, while still avoiding eating the skins of chicken and pork.

A month and a half later, Ron called his mom from the United States. She told him that the rashes had mostly gone, and the swelling in her feet was subsiding. Three months later, her rashes were completely gone. Her feet had gone back to their normal size, and only a few hairs fell off when she was combing.

Ron was happy. He had saved his mom.

Example 10 Itchy skin

Ron did have some experience with itchy and dry skin.

His wife's sister, Claire, immigrated to the United States fifteen years ago. A few years later, she married Angus, a gentleman from China. One day, Claire was visiting, and was telling Ron's wife that Angus recently had rashes all over his body. Angus had gone to see his family doctor, who had then prescribed some medicated skin cream. The cream came in a very small bottle (80 ml), and cost over $30. As they were not particularly well off, they considered that expensive.

By chance, Ron overheard the conversation. A few months earlier, before Angus and Claire moved into their current apartment, they came to stay with Ron and his wife for a couple of weeks. Ron remembered that after Angus finished taking a bath, the bathroom was so steamy that it looked like a sauna. Now that Angus had rashes, Ron could figure out what the problem was.

He told Claire to tell Angus not to use very hot water for a shower. He should use lukewarm water, and avoid using soap for a while. Hot water, as well as soap could remove the natural oil protecting the skin.

Angus followed the advice and in about two months later, the rashes were all gone.

It is important that we monitor our daily activities and surroundings, as we are the ones that suffer when adversity falls upon us.

Do I get a stomachache after drinking the leftover soup in the fridge? Do I get sore throat from eating deep-fried food? Does my mouth feel dry after a meal in the restaurant, where monosodium glutamate (MSG) is quite often used as a flavor enhancer? Do I feel drowsy after using the liquid cleaner to clean the bathtub? Am I allergic to the new blanket that I just bought?

Medical doctors do not know our daily habits. Since we are the ones who are exposed to all these irritants and diseases, we should take note of what we eat, drink and breathe in.

Example 11 Mutual fund price

The percentage gain of a mutual fund is sometimes quoted in terms of 1-year return and 2-year return. If the 1-year return is given as 20%, and the 2-year return as 5%, it looks as if the fund is profitable all the time. But the fact is that it actually lost approximately 10% for the first year of the 2-year period. This can be roughly estimated as $5\% \times 2 - 20\% = -10\%$. (Using a more accurate calculation, the fund actually lost 8.3% for the first year of the 2-year period.) But this information is quite often not revealed, and the investor may not be aware that the fund is actually quite volatile.

Statisticians have a joke about their profession. The joke goes like this: "Statistics are like a bikini. What they reveal is suggestive, but what they conceal is vital". Some of the hidden information may be more important than the information being unveiled. What is the small print in the contract? Are there any warranties for the products that we purchase?

While some information is hidden but still can be extracted, we are sometimes faced with situations where we have no information at all. Can we do anything about this? We will discuss this in the next section.

3.1.4 *No information*

Occasionally, we will run into circumstances where no information is available. Neither time, nor resource, at hand allows us the luxury to search for any relevant information. And we need to make a decision or judgment then and there. Is there any experience that we can draw upon? Fortunately, sometimes there is. We may be able to rely on general principles that we or other people have concluded by way of induction from various observations. From the general principle, we can deduct our action for the specific situation we are facing. We will discuss more about this when we come to the chapter on Induction and Deduction. For the time being, we will take a look at an example.

Example 12 Smoked meat sandwich

There is a restaurant in Montreal that is famous for its smoked meat sandwiches. The restaurant can only seat about fifty people. It does not take reservation. Customers usually have to line up for about an hour to get in, and they may have to share the table with one or two more groups of customers.

It was Christmas; father, mother, and their twenty-two year old daughter went to Montreal to do some sightseeing. They were told about this restaurant, and decided to go there for lunch. After waiting for about an hour outside, they were led inside and seated by a waiter. They looked at the menu. Smoked meat sandwich cost $4.95. They thought they would order three sandwiches. However, the daughter noticed that one could also order a large plate of smoked meat for $9.95. It would come with bread, and one can make one's own sandwiches. Making one's sandwich meant spreading mustard on two pieces of bread, and then putting the smoked meat in between the pieces of bread. As $9.95 was about the same as the cost of two smoked meat sandwiches which would cost $9.90, the question was whether ordering a large plate would be a better deal than ordering two sandwiches.

As this was their first time in the restaurant, they would not know whether a large plate would have twice as much smoked meat as the meat in a sandwich. Nevertheless, they could draw on two general economic principles. Firstly, it is general true that the more one buys, the cheaper per unit the merchandise is. For example, it is usually cheaper per toilet roll if one purchases a 12-

roll package than a 6-roll package. Secondly, merchandise is usually cheaper if the customer has to do some preparation himself or herself to get to the final product. Thus, a home-cooked meal is usually cheaper than a restaurant meal, providing the ingredients are the same.

In the current situation, based on the first principle, the daughter figured that the $9.95 plate should contain much more smoked meat than the meat in two sandwiches. Furthermore, based on the second principle, as the customers had to make their own sandwiches and that cut down on the manufacturing time of the restaurant kitchen staff, the $9.95 plate should again contain much more smoked meat than the meat in two sandwiches. The daughter eventually decided to order one sandwich as well as one large plate of smoked meat with bread for the three of them. When the order arrived, they could see that the large plate contained approximately 25% more than two sandwiches, both in the smoked meat, and the pieces of bread.

In this case, we can see that even though the daughter did not have any information about the quantities in different orders in that restaurant, she was right in her decision by applying deduction from various general principles. In other words, as no external information is available, she attempts to make use of the internal information that is already stored in her mind.

In the above example, the person involved knows that she does not know the information, so she knows that she has to make up for it somehow. However, there are situations that the person affected simply does not know that the information exists to begin with, so she does not know what she is missing at all.

3.1.5 *Unaware information*

There is so much information in the world that it is definitely impossible for us to know everything. So, when we are facing a problem, we would search for information that we think is relevant to the problem at hand. However, there can be information that is relevant, but we are completely unaware of its existence. In this case, we will not be able to solve the problem, or at best, come up with a less favorable solution.

Our path of knowledge can be expressed in terms of a table of "don't know" and "know" as follows:

	don't know	know
don't know	don't know we don't know	don't know we know
know	know we don't know	know we know

Our learning process starts out with a blank slate. We begin with the element "don't know we don't know", and then proceeds counter-clockwise to eventually arrive at the element of "don't know we know". An example will clarify this table. We will take the learning of riding a bicycle. When we are first born, we don't know that we don't know how to ride a bicycle, as we have not even seen a bicycle to begin with. As we grow older, we can see that other people can ride bikes, and we know that we don't know how to ride bikes. So, we try to learn, and eventually master the skill. As a result, we know that we know. As time goes by, riding a bike becomes a second nature to us and we completely forget that we know how to ride a bike. That is the stage when we get to the point of "don't know we know".

When we know that we don't know, we will look for the information. When we don't know that we don't know, we do not know what to look for, or that we need to look for at all. One of the unfavorable situations in problem solving is "don't know we don't know" that certain information exists. As such, we will not even search for the information. Let us take a look at an example.

Example 13 Air travel

In the year 1996, Lilian was living in Toronto, Canada. She had to fly to Tokyo, and then take a train to Sendai for a conference. After the conference, she wanted to fly to Hong Kong to visit a friend. So she bought a return ticket from Toronto to Tokyo at a cost of $1,300, and then another return ticket from Tokyo to Hong Kong at a cost of $700.

At the conference, she met Heather who was also from Toronto. As it happened, Heather was also flying to Hong Kong after the conference to visit her sister. Heather told Lilian that she simply bought a return ticket from Toronto to Hong Kong, with a stopover in Tokyo. All she paid was $1,200. That price was even cheaper than the return ticket from Toronto to Tokyo that Lilian paid. Lilian was simply unaware that she could have bought a ticket the way as Heather did. Not aware of such useful information cost Lilian much more money.

There is not much we can do about not being aware of certain information. However, keeping our eyes open to our surroundings would help. As well, talking to other people is definitely beneficial. Other people sometimes get things done completely different from what we can even dream of, and that quite often provides us with ideas that how some problems can be better solved.

3.1.6 *Evidence-based information*

Evidence-based Medicine (EBM) was developed in the 1990's. The basic premise is to discard the declaration of authorities, and seek the facts from systematic observation in patients. New evidence in clinical research can challenge and refute previously accepted diagnostic examination and treatments, as well as replacing them with more reliable and safer therapies. This approach would lead to healthcare professionals using the best research evidence in their everyday practice.

For example, based on clinical studies, EBM supported the benefit of steroids in lessening respiratory distress in premature babies, in spite of the age-old belief that steroids could be damaging.

This evidence-based approach has since been employed in information gathering in other disciples, e.g., education, social work, marketing, management, and financial market trading.

Thus, we should always try to find out whether any information is accurate, and not based on hearsay. All this knowledge can be stored and sorted out in our mind. When required, all we need to do is to make use of this internal information to tackle the problem situation.

3.2 Internal information

Having a reserve of data and facts is essential when dealing with everyday problems. Unfortunately, sometimes, even though a person already has or has been given the correct information, because of pride or some other emotional reasons, he or she refuses to believe it, as the following two examples will show.

(A) Emotional

3.2.1 *Self-denied information*

Example 14 Grammatical errors

Meg works as a manager for a company. Quite often, she has to write memos. Occasionally, before she sends the memos out, she will bring the drafts home, and asks her husband, Tom, for comments. Tom notices that she has made quite a number of grammatical errors, and has pointed that out to her several times. However, Meg insists that the grammatical errors are insignificant, and it is the flow of the content that is important. Eventually, Tom gives up, and would not comment on her grammatical errors, even though sometimes he thinks that some of the errors are so serious that they would make the content ambiguous.

One day, Meg came home, and told Tom that one of her colleagues said that her writing needed improvement, and she was wondering why. Knowing that she would not like to accept her faults, Tom did not make any remarks.

No one likes to be criticized, but it is important to accept the facts and admit one's mistakes. One should then make changes and improvements to one's tasks.

While some people reject information that they do not like, some people unjustifiably choose to put emphasis on certain pieces of information. They are biased from the very beginning, and are very selective in the information they pick, as the following example will show.

3.2.2 *Biased information*

Example 15 Renovation

Mary is an interior designer. A friend of hers bought a house that needed to be renovated, and had asked her to do the interior decoration. Mary wanted the interior of the house to look attractive. However, she would ignore safety standards and override other contractors, if she did not think their proposals fit her ideals.

For all the home products she picked for the house, her main concern was whether they looked attractive, not whether they were effective or reliable. She chose a fancy-looking door lock, against the advice of the locksmith who did not think it was dependable. As a consequence, a year later, it was necessary to change the door lock, as there was difficulty opening the lock with the key.

In addition, she picked toilets in an upscale model and design. However, the homeowner later found out that the toilet handle must be held down to complete the flushing action. The plumber was not able to adjust the lever inside the toilet tank to fix the problem.

Mary prefers to think that she is always right. She chooses information that fits her liking, and ignores other people's recommendations. However, when facing a problem, we always should have an open mind, and should consider all relevant information. We definitely should not let our prejudice and emotion take the better part of us.

(B) Unemotional

Assuming we do not let our feelings control our judgment, and are quite rational, that still does not mean that we can see the connection of different concepts. Knowing certain information does not necessarily imply that one knows how to apply that information to existing problems, as one may not see the relation. Let us take a look at the following example.

3.2.3 *Unexploited information*

Example 16 Activity-based costing

In March, 2008, Willie, an accountant working for the Canadian Federal Government, was moved to a new sector specializing in activity-based accounting. She was quite excited about her new job. Her husband Peter, a scientist, was self-employed and worked at home. Being a non-accountant, he asked her what exactly was activity-based accounting. Willie then explained.

Activity-based costing is a cost accounting method that was developed in the late 1980's. Traditional cost accounting arbitrarily adds a certain percentage of the expenses to the direct cost to allow for the indirect costs like rent, taxes, telephone bills etc. However, as manufacturing a product or providing a service becomes more and more complicated, this traditional method cannot provide an accurate measurement of the actual cost. Activity-based costing identifies, describes, and assigns costs to each activity that produces a product or service, and is now considered to be a more accurate method of costing.

A few weeks later, Willie called Peter from her office at nine in the morning. She just got in. However, she had forgotten her monthly parking pass at home, and therefore, could not get into the government-parking garage where she regularly parked her car. Instead, she had just parked her car at the pay-and-display city parking across the street. (Pay-and-display is where one purchases a parking ticket from a machine to allow the car to park to a certain time. The ticket is then displayed on the dashboard of the car.) Could Peter drive over and drop off her parking pass at her office and be there in half-an-hour, so that she could park her car in the government-parking garage after? Peter reluctantly agreed.

However, as Peter was driving to drop off the parking pass. He was wondering whether the trip made economic sense. Parking in that city-parking lot for half-an-hour costs $2, which, presumably, was what Willie had paid. But parking from 7 am to 5 pm only costs a maximum of $10. 5 pm was when Willie got off work. It would take Peter 20 minutes to drive one-way to Willie's office, and 40 minutes to drive both ways. The whole trip would cost about $7 gas for his car. Taking into account the wear-and-tear and depreciation of his car, as well as his time, the trip was not worth it judging from activity-based accounting. Willie should have paid $10 at the city parking and left her car there until 5 pm, when she got off work. She had failed to see the hidden cost of Peter's driving to her office. She had not related her professional knowledge to an everyday problem.

Noticeably, it is not good enough just to simply store the information in one's mind. One needs to be able to make use of the information and apply it to the problem on hand. One should be able to see the relationship between one's expert knowledge and the new and unfamiliar situations that one encounters everyday.

On the other end of the spectrum, there exists knowledge in our mind that we seldom used, or not very familiar with. However, there is no particular reason why it cannot be exploited. If one can make use of this peripheral information, i.e., information that is not central to one's expertise, one would have a much larger arsenal of tools to work with. Occasionally, the laymen and the amateurs can beat the professionals in doing the tasks better, as the next example will show.

3.2.4 *Peripheral information*

Example 17 Bathroom sink faucets

The Jones had just moved into a 10-year-old two-storey house. The house has a two piece bathroom (toilet and sink) on the ground floor.

A couple of days later, Mr. Jone found that the two-handle (one for hot water and one for cold water) centerset faucets at the sink were loose. Centerset faucets are faucets where the spout and handles are attached to one base. He looked under the sink, and found that there were water stains as well as rust at the bottom of the sink cabinet. Water had been leaking under the faucet base into the sink cabinet. The water also rusted the two metal nuts that secured the hot and cold water faucet tailpieces to the underside of the sink.

The sink top was made of marble and had three mounting holes. The middle hole allowed a pop-up assembly to go through for controlling the pop-up drain. The two other holes allowed the hot and cold water supply lines to be connected to the two faucet tailpieces that were inserted through the holes. The plumber who installed the faucets must have known that there could not be much friction between a metal faucet base and the marble top, and tightening the two metal nuts (that were screwed onto the threaded faucet tailpieces) against the underside of the marble top would not have secured the whole centerset faucets. The whole faucet assembly would have wobbled sooner or later as the diameter of each faucet tailpiece was smaller than the diameter of the mounting hole that the tailpiece was inserted into. So, the plumber simply wrapped some paper respectively around the hot and cold water faucet tailpieces so that the paper would fill the empty space in the mounting holes.

That, of course, could not have kept the faucets secured for long. After a while, water leaking through the faucet base would have made the paper soggy, as well as causing the metal nuts to get rusty, and eventually bringing about the whole faucet assembly to be wobbly against the sink top. In addition, the leaking water gave rise to water and rusty stains at the bottom of the sink cabinet.

To fix the problem, Mr. Jone first went to a home hardware store to purchase two plastic mounting nuts and two O-rings. (Rubber gaskets probably could be used instead of O-rings). He removed the faucets, and threw away the two rusty metal mounting nuts as well as the papers that the plumber used to stuff the two holes. He then put the two O-rings between the marble sink top and respectively the hot and cold water faucet bottom plates. This would make the whole faucet assembly stable when the two plastic mounting nuts were used later to tighten the faucet tailpieces against the underside of the sink. Finally, he sealed the perimeter of the faucet base plate with silicone rubber caulking so that water will not leak into the sink cabinet. After he finished, the faucets were not wobbly as they were before, and water did not leak into the sink cabinet. He believed he had done a better job than the plumber who installed the faucets in the first place.

The important element is that we should make the most of whatever information that we have already stored in our brain. Some knowledge that we barely know or familiar with can be exploited to our advantage. If the accumulated information is not enough, we should search for other relevant knowledge.

We should also be careful not to have a built-in-assumption that certain information must be correct. It may sometimes take quite some observation and specific hypothesis to cast doubt on certain existing information. The hypothesis would tentatively explain our observation or any deviance from the norm. However, it needs to be tested with further observation in order to be confirmed or disproved. We will take a look at hypothesis in the next chapter.

Chapter 4

Hypothesis

In scientific discipline, a hypothesis is a set of propositions set forth to explain the occurrence of certain phenomena. In daily language, a hypothesis can be interpreted as an assumption or guess. In this book, we employ both these definitions. Within the context of the first definition, we search for an explanation of why the problem occurs to begin with. Within the context of the second definition, we look for a plausible solution to the problem.

For some problems, it is significant to be able to explain why certain events happen (e.g., in some medical problems). For other problems, we can ignore what causes the events to happen, and go straight to solving whatever problems have arisen (again, e.g., in some medical problems).

Depending on the nature of the problem, both approaches of hypothesizing are useful. Sometimes, one approach is superior to the other, and at other times, vice versa. Let us first take a look at some examples why in certain situations, it is important to understand why certain phenomena occurs.

Example 1 Visiting cats

A couple moved to a house in a different neighbourhood. The kitchen was at the back of the house, facing a backyard with lots of flowers. They could sit in the eat-in area of the kitchen, and look at the backyard through an all-glass patio-door.

A few days after they moved in, they were having lunch at the eat-in area. As the wife turned her head to look at the backyard, she saw a cat outside the patio-door staring at her. As it happened, she had a cat phobia, and was shocked in seeing the cat. Fortunately, the cat left one minute later. For the next two weeks, there were different cats coming over to their patio-door, and that scared the heck out of her.

The couple discussed several plausible ways to bar the cats from coming. The backyard had been fenced off only with hedges, and the cats could easily go through them. If they had to stop the cats from coming, they would have to build wooden fences around the backyard. That would cost thousands of dollars. Alternatively, they wondered whether there might be some ultrasonic devices that would drive the cats away. They mulled over several suggestions for the next few days, and still had not come up with a cost-effective solution.

A couple of days later, the wife suddenly remembered that the previous owner had a cat. She recalled seeing the cat for a brief moment when she and her husband came to look at the house before they decided to buy it. The cats that came to the patio-door must have been coming over to ask their friend to go out and play. Once she figured that out, practically nothing needed to get done. Cats are smart animals. They would soon realize that their friend had moved, and would stop coming to look for him. As it happened, a couple of weeks later, no more cats came to the patio-door.

In this particular case, once the cause of the problem was perceived, no action needed to be taken.

Example 2 Skin rashes

Mary was born in Macau. She had four brothers and two sisters. Her mother died when she was five years old. Her father was those kinds of men that did not seem to care much about their kids. So, after the mother died, Mary's grandmother took up the responsibility of bringing up the children.

When Mary was a teenager, she had rashes all over her body, including her legs. Her grandmother took her to see the doctor, who prescribed some medicated cream. That did not seem to help too much. For the next few years, Mary tried both Western and Chinese medications, but the rashes would not go away. Once, her grandmother heard of a concoction where some Chinese herbs would be mixed in with honey, and someone said that it would cure rashes. She prepared some and spread it on Mary's body. The concoction was sticky, and Mary hated it. In any case, it did not do anything.

As a teenager, Mary was very self-conscious about her rashes, especially when she was wearing skirts. She would think that was probably why she did not get many dates. After high school, she went to England to study in a college. During the two years that she lived in England, amazingly, she did not have any rashes.

After finishing her study in England, she came back to Macau. By then, her family had moved to another house. Her rashes came back, though not as serious as it was before. One of her friends suggested that it might be the drinking water in England that might have made a difference, but she did not think that was the reason.

A few weeks later, it suddenly dawned on her that it might have something to do with the washing machine. She remembered that before she went to England, her grandmother sometimes complained that water was leaking out of the old washing machine. Now that they had moved to another house, they had bought a new washing machine, and her rashes were not as severe. Could it be possible that some laundry detergent was not rinsed off completely and still clung to her clothing after the washing cycle had finished, and she was allergic to the detergent? From then on, she double-rinsed her clothing, i.e., after the washing cycle had finished; she turned the washing machine dial to the rinse cycle again, and re-rinsed her clothing once more.

That seemed to have solved the problem. Her rashes slowly disappeared, and in a month, she did not have skin rashes any more. After suffering for seven years, she finally found out the reason why she had rashes to begin with.

It should be noted that the information was there all the time. Unfortunately, no one in her household had come up with the hypothesis to explain the cause of the problem. Once the explanation was found, the problem was easily solved.

Nevertheless, in some other problems, we do not have to understand their causes, we can take a short cut, and go directly to find a solution, as the next two examples will show.

Example 3 Bladder control

Chee is a smart lady. She finished high school, and worked as an elementary school teacher for a number of years. She retired early, and spent her time watching the stock market. She did not know how to use the computer, and barely knew how to use the calculator. So, she would write the market indices and stock prices in a little black book. She watched the stock prices going up and down, and would buy low and sell high. Interestingly, she consistently made money out of the market.

She used to practice Tai Chi (a form of Chinese shadow boxing), which can be considered as a combination of a moving form of yoga and meditation. In her late seventies, she began to find it difficult to practice some of the movements in Tai Chi, so she invented her own exercises. Every morning, she would spend an hour doing her own exercises in a park near her apartment. She also watched her diet, and maintained a healthy life style.

About seven years ago, in her early eighties, she started having bladder control problems, and occasionally wetted her pants. This urinary incontinence problem was not an uncommon dilemma for elderly people. So she went to see her family doctor, and was told that there was nothing she could do. She just had to wear senior diapers for the rest of her life.

Undeterred, she invented her own exercise to control her bladder. She would stand on the ground with her feet about half a metre wide. She put her hands on her belly, took a deep breath, and held her breath until she could not hold any longer, and then breathed out. She would repeat this breathing exercise fifteen times. She practiced this twice a day, once in the morning, and once in the afternoon. One week later, she had her bladder under control. She continued doing this exercise daily, and she did not have any leakage problem ever since.

Chee did not attempt to understand the cause of her problem. It would have been too complicated for her to comprehend. Instead, she tried to come up with a solution. And it worked.

Example 4 Common cold

David caught a cold on the average once a year. He would have a sore throat and then a runny nose. Sometimes it got so bad that he could barely breathe. It usually lasted for about four to six weeks, and the cold would run its course and cure itself. In his twenties, it bothered him but it was bearable. In his thirties, he found having a cold more and more unbearable. It also affected his work efficiency when he caught a cold. Once, his throat was so sore that he eventually had to go see a doctor. The doctor prescribed anti-biotic, which cut short his discomfort. From then on, once he started getting a cold, he would go to see the doctor, and asked him to prescribe anti-biotic to destroy the bacteria. That would cut down the duration of having a cold to about three weeks, which was quite some improvement.

One day, he heard from a nurse, a friend of a friend of his, that taking antibiotic too often was not good, as bacteria quite often develop tolerance and resistance to the medication over time, making them difficult to be destroyed in the future. He then started to think whether there was any way to avoid catching a cold to begin with. Initial symptoms varied from person to person. Some people started having a congested nose; other people might start having a sore throat. For him, he always started having a sore throat. The bacteria would eventually travel up to his nose, and he would have a runny nose.

To avoid getting a cold, he just would have to nip the sore throat in the bud. So, he came up with an idea. At the slightest sign of a sore throat, he would try sucking sugarless candy continuously. (He figured that sugared candy would probably do as well, but sugar was bad for the teeth.) Not only was the saliva generated by sucking a candy soothe his throat, he believed that the saliva might serve as an antiseptic killing off some of the bacteria. The sore throat would usually subside in a few days, and would not lead to a runny nose. And even when he occasionally gets a runny nose, it is not as severe as it was before and will last for about a week. The idea of sucking candy thus seems to work, and for the past twelve years, he had caught the cold only once.

Again, in this particular case, a solution could be found without understanding the causes of the problem.

As we can see from the above examples, it is wise to try to spend some time to think and come up with a hypothesis as early as possible, rather than doing nothing or spending a long time in performing observations or collecting information. Coming up with a hypothesis quickly can help us plot our next path or make our subsequent decisions, as the following examples will show.

Example 5 Restaurant

When Ricky was a university student, budget was tight, and he rarely ate out in a restaurant. Once, a friend of his was having a birthday party, and a few of them went to dinner in a restaurant well known for its good food. There was a long line-up at the door, and they did not want to wait for an hour to get a table. Instead, they went to the restaurant next door. Fortunately, it was only one-third full. They sat down and ordered their food. The food came and they started eating it. And then they found out why the restaurant was not full to begin with. The food was so bad that it would be much better to eat at a fast food restaurant, where the food quality and price would be more reasonable, and they would not have to wait. When Ricky told his friend Steve later on about the lousy restaurant, Steve told him he had a similar experience. He and his girlfriend were traveling in Britain. They were hungry and went to a restaurant close to where they were sightseeing. All the tables in the restaurant were neatly set, but there was no other customer eating there. As they were reading the menu, they could see dust on the cups and plates. And they wondered how many people had been eating there for the last month or so.

Ricky quickly forged a hypothesis — that a restaurant is mediocre if it is less than half-full at mealtime. From then on, if he is not familiar with a restaurant, he will go inside and take a look first. If there is no person or only a couple of people eating there, especially during mealtime, he will just walk out, and try somewhere else. When he is travelling in a foreign country, if he does not see any local people eating in that restaurant, he would think twice before eating there.

Of course, one's hypothesis may not be correct. If it is shown later on that the original hypothesis is not correct, then one should reevaluate it or discard it and quickly come up with another hypothesis. Let us take a look at the following examples.

Example 6 Flies in the house

John lived in Toronto, Canada. He moved to a new house a few months ago. One Saturday, he saw a few houseflies flying in the house. He hated flies. Flies eat food from the garbage, which can contain germs. They can regurgitate saliva on our food, transferring some of the germs from the garbage. In addition, they carry bacteria on the outside of their bodies, especially on their sticky feet. Everytime they walk on our food, they leave some of the bacteria behind.

John quickly got a fly swatter and killed the flies. A few minutes later, he saw several flies flying in the house again. Again he killed them. This went on and on for an hour. Within that hour, he killed about twenty flies. He figured that they must have come in through some cracks in the house, as all the windows in the house were closed. It was a hot day and 27 degrees Centigrade outside. John had his central air-conditioning on. Could it be possible that the flies want to come in and enjoy the air-conditioning?

He knew that his next door neighbour had his central air-conditioning on also, as he could hear it running from his house. He ran into his neighbour the next day, and asked him whether he had any flies flying in his house the day before. His neighbour said no.

A week later, he saw flies flying in his house again. Then, he realized that for both times he was boiling soup for more than an hour, and he had his exhaust fan over the stovetop on. The flies might have smelt the smell from the soup, and came in through the exhaust. He watched, and saw a fly creeping in through a crack in the exhaust hood.

From then on, when he was boiling soup, he did not turn on his exhaust fan. Instead, he put a large dome-shaped aluminum cover on top of the lid of the cooking pot. The steam from the pot would condense onto the cover, and drip onto the stovetop, and he would wipe off the water later. This way, the steam would not escape into the air and make the house too humid. If the kitchen smelt, he could use any of the commercial odor-removal products in the market. Since then, he did not see any flies coming into the house any more.

Later on in the winter, even though there was no fly outside, he still did not use the exhaust as he found the dome cover method particularly useful. The heat stored in the steam from the boiling soup, instead of escaping through the exhaust, would stay inside the house, thus increasing the energy efficiency. This, of course, also contributes to reducing global warming. The dome cover is necessary to make the house less humid. Winter in Toronto can get very cold. If the house is too humid, water vapor inside the house can condense onto the cold glass windows, and turn into ice. When the sun shines on the window, the ice melts into water. The water, if left unwiped, can damage the paint as well as the wood of the windowsill.

Example 7 Missing sunglasses

Teresa and her husband live in New York City. One summer, they flew to San Francisco for some sightseeing. They spent a week there and had a good time.

On the last day of their trip, they checked out of their hotel at ten in the morning. While they were in the parking lot of the hotel, Teresa suddenly realized that her pair of sunglasses was missing. It was a pair of designer glasses, and cost about $300. As she hypothesized that she might have left them at the restaurant where they had lunch the day before, she called the restaurant, but was told that no one had found any sunglasses.

A couple of days after they flew back to New York City, Teresa hypothesized that she might have left her sunglasses in the hotel room. So, she called the hotel, and was told by the housekeeping that they did pick up her sunglasses, and could send them back to her if she paid for the postage. She agreed.

The sunglasses arrived a few days later. After Teresa opened the flimsy envelope, she found that the sunglasses were broken into two pieces, right in the middle, and therefore completely useless. The housekeeping in the hotel did not package the sunglasses in a sturdy envelope, and they were broken in transit.

In hindsight, Teresa should try to come up with another hypothesis right after the restaurant told her that they did not have her sunglasses. Did she remember seeing or holding her sunglasses after she left the restaurant? When did she last see her sunglasses? If she had then thought that she might have forgotten them in the hotel room, she could have easily gone back up to the room to check as she was still in the hotel area.

As we see in the above two examples, if the first hypothesis is not correct, we should quickly come up with a second hypothesis to explain the incident, and hopefully solve the problem. But then, is there any method that will allow us to pick the correct hypothesis, or increase our chances of finding the right hypothesis? Let us first take a look at what abduction or abductive reasoning is all about.

4.1 Abduction

Abduction is a method of reasoning applied in the scientific world where a hypothesis is chosen to best interpret a phenomenon. It attempts to provide a theory explaining the causal relationship among the facts. If hypothesis H explains a set of facts better than other proposed hypotheses, then H will be chosen as probably the correct hypothesis. Thus, abduction can be considered to consist of two operations: the formation, and selection of plausible hypotheses. This kind of reasoning has been implemented in artificial intelligence for various tasks, e.g., medical diagnosis, automatic fault detection, and speech recognition.

Several hypotheses can be deduced, but eventually we have to pick one that we think would most likely explain our observations. The hypothesis chosen has to be consistent with existing theories. This, of course, does not mean that existing theories are necessarily correct, as theories should be modified or discarded if they do not agree with new experimental evidences. However, to make things less complicated for the time being, let us assume that the theories are correct to begin with. The hypothesis selected should be compatible with the theories and should better explain the observations than alternative hypotheses. We should also consider the cost of being wrong and the benefit of being right. In Statistics, accepting a hypothesis when it is actually incorrect is called a Type II error, and rejecting it when it is indeed right is called a Type I error.

To come up with the correct hypothesis for an everyday problem, there are two factors that can be beneficial. Firstly, a general knowledge of various disciplines would be useful, and a basic knowledge of some of the science subjects (e.g., Biology, Physics, and Chemistry) would be helpful. Secondly, we should train ourselves to quickly see the relations among various concepts (See chapter on Relation). Understandably, our knowledge is always limited. Thus, it is important that we exploit all the knowledge that we possess, including the information that we may not be too familiar with. We should then try to combine different notions in our mind as the various associations can multiply and extrapolate what we can usually come up with.

A hypothesis needs confirmation to be shown that it is indeed correct. Experiments need to be performed to verify the hypothesis. We will discuss experiment in the next chapter. However, before we do that, we should point out that we may not be able to apply experimentation (in the strict scientific sense) on quite a number of everyday problems. Nevertheless, hypothesizing does allow us to solve some of the problems. This is particularly valid when we encounter unfamiliar situations. We cannot be knowledgeable in all areas. People familiar with that particular area may find the problem trivial. But since we may not be accustomed to certain environments, we just have to make wild guesses to problems that are unusual to us. Occasionally, some of the crazy ideas do work. Let us take a look at the following examples.

4.2 Wild conjectures

Example 8 Lousy dish

A family of four went to a restaurant. They would like to order a set dinner for four. The mother also saw a seasonal fish dish as one of the recommended items on a one-page special menu. Before ordering the fish dish, she checked with the waitress, and asked her whether it was any good. The waitress thought highly of the dish, as she said she and her husband tried it last week, and it tasted excellent. So, the mother ordered the fish dish, together with the set dinner for four.

When the fish dish came, the family found that not only did the fish tasted lousy, it was way overcooked. But why was the fish tasted so lousily from what the waitress had experienced? The father came up with a hypothesis. That was a small restaurant, and chances were, there were only two chefs, one good one and one mediocre. When the order was taken to the kitchen, someone saw that one of the orders was a set dinner, and gave the whole order to the mediocre chef, knowing that a set dinner did not require too much skill to cook, and was actually quite standardized. The same cook made the fish dish, and did a poor job on it. The fish dish should have been prepared by the good cook as; in general, seafood dish requires better skill.

A few weeks later, the family went to the same restaurant again. They ordered a set dinner for four, as well as a lobster dish. But this time, the husband asked his wife to put the two orders in separately, thinking that the set dinner order would be given to the mediocre chef, and the lobster dish order would be given to the good chef. The wife followed her husband's suggestion, and put in the lobster order five minutes after the set dinner order. When the lobster dish came, it turned out to taste very well, and was cooked just right.

Example 9 Tour bus

In May 2006, Ben and Janet joined a multi-destination China tour. One of the destinations was Yellow Mountain, the most beautiful mountain in China.

One morning after breakfast, the tour group checked out of the hotel, and would be driven to the Yellow Mountain in an air-conditioned tour bus. The trip would take about one-and-a-half hour. The bus was only about three-quarter full, with most of the passengers sitting in the front, leaving the back part of the bus half-empty.

The air conditioning in the bus was running full-blast. Passengers sitting in the front, where it was quite crowded, were feeling quite comfortable. However, passengers sitting at the back, where it was much less crowded, were feeling cold. As sweaters and jackets had been packed away and stored beneath the bus in the undercarriage luggage compartment, they were not easily accessible. The obvious solution would be to try to turn off the air ventilation control nozzles that were mounted below the overhead luggage rack and above the heads of the passengers. (There was one nozzle for every two passengers.) The passengers sitting at the back soon found out that no matter how they turned the nozzles, clockwise or counter-clockwise, the airflow was still the same. It did not seem as if the airflow could be adjusted at all. After a few moments of adjusting, they simply gave up.

Ben and Janet were sitting at the last but three rows. They were also feeling the drift of the cold air from their nozzle above their heads. Janet asked Ben whether he could do something about it. Ben did not think so, but he tried anyway. He too, soon found out that the cold air could not be shut off, nor reduced by turning the nozzle.

Ben figured that those individual shut-off valves were never installed to begin with, either intentionally or unintentionally. If so, what else could be done? He then realized that all he needed to do was to block off the cold air drift. Being an observant person, he noticed the curtain hanging over the window, and he got an idea. He lifted the curtain to cover the nozzle, and held the bottom of the curtain over at the overhead luggage rack, putting a carry-on bag on top of the bottom of the curtain so that the curtain would not fall down. The curtain blocked out the drift of cold air coming from the nozzle, and Ben and Janet did not feel cold anymore. Other passengers saw what Ben did, and soon followed his idea.

There are a number of situations in which we do not know the inner workings, nor could control them even if we did. However, by hypothesizing, we may come up with an idea of how the problem can be solved. The next example shows how someone discovered that a game of chance was not that random after all and how he directed his son to win the game. It will demonstrate that keen observation and wild hypothesis can be fun and rewarding.

Example 10 Water squirting game

The author of "Science of Financial Market Trading" (World Scientific, 2003) described his own experience with a water squirting game in an amusement park:

"About ten years ago, my whole family was visiting Hong Kong. I went with my five-year-old son Anthony to an amusement park. One of the games in the park was a water squirting competition that had ten seats. Each participant had a water pistol. Water going into all the pistols would be started by an operator. Each person would aim the water at a wooden clown's mouth, which was about one metre right in front of each pistol. As water was shot into its mouth, a ball would rise up a tube, which was connected to the mouth. The first person that managed to raise the ball to the top could win a prize. We stood there watching several games. The people who sat on the leftmost side always won. I hypothesized that water must be piping in from the left side and distributed to all the water pistols, thus the water pressure from the leftmost side was the highest, contributing to the people sitting there always winning. I mentioned this conjecture to Anthony. However, we did not stay to play any game.

Back in Ottawa (Canada) and a year later, we went to an annual amusement exhibition and saw a similar game. The choice of prizes included stuffed lobsters. This was the first year that the exhibition had stuffed lobsters and they were cute. My children Angela and Anthony would like to have them. Anthony immediately went to the leftmost seat and started playing. He lost. At that point, I told Anthony, "Stop playing and let me watch for a while". For the next few games, people sitting at the centre seats always won. What happened was, this game had nineteen sets, much more seats than the one in Hong Kong. I figured that the water must be piping in from the centre and distributed to the water pistols on both sides. I then asked Anthony to sit in the centre seat. He won three out of four games. He got three small lobsters, and he could exchange two of them for a large one. The kids were happy. So was I. I have found that an apparently random game was not so random after all."

Occasionally, wild ideas and farfetched hypotheses can solve some of the apparently unsolvable problems, or revolutionize the existing landscape. One of the bold hypotheses in science of all time is as follows:

A number of observers are moving at uniform speed with respect to each other and to a light source, and if each observer measures the speed of light coming from the source, they will all obtain the same value.

This hypothesis is counter-intuitive to the point of the unimaginable. It completely contradicts classical physics. And it is not surprising that it is proposed by none other than Einstein. Its strange content sets the stage for the Special Theory of Relativity, contributing to a revolution in physics.

4.3 **Albert Einstein** (1879–1955)

The Special Theory of Relativity was developed by Einstein in his spare time while he was working full time as a Technical Expert, Third Class at the Swiss Patent Office between 1902 and 1905. Before we find out how Einstein came up with the hypothesis of the constant speed of light, we need to take a look at the two fundamental principles that Einstein bases it on.

The first principle that Einstein asserts is that all laws of physics take the same form in a vehicle, whether that vehicle is at rest or in uniform motion. This simply implies that no experiment of any kind can detect absolute rest or uniform motion. This he calls the principle of relativity, and is actually a modification of a deduced principle from Newton's laws as stated in Newton's *Principia Mathematica* published in 1687.

The second principle that he claims is that in empty space, light travels with a definite speed c that is independent of the motion of its source. (We will take c to be in Kilometres per second in the following discussion. However, its unit is not important for the argument.)

Both principles look innocent, but they do have significant consequence.

Imagine a vehicle with a lamp at its centre. Consider first that it is in an absolute rest position. At a certain point in time, the lamp is flashed for an instant, sending out pulses of light to the left and to the right. The speeds of light are measured at the left side and the right sides, and it is found that the value is c for both cases.

Now consider that the vehicle is moving at a uniform speed of 10,000 Kilometres per second to the right. At a certain point in time, the lamp is flashed, again sending out pulses of light to the left and to the right. Two experimenters, A and B, stand inside the vehicle, with A on the right and B on the left. Both measure the speed of the pulses. Now the question is: what values of the speeds of the pulses relative to A and B would they obtain?

According to Einstein's second principle, the speeds of the pulses of light are independent of the motions of their sources. Now, because of the uniform motion of the vehicle to the right, one would expect A to find his rightward-moving light pulse traveling relative to him with a speed of (c – 10,000) Kilometres per second, and B to find his leftward-moving light pulse traveling relative to him with a speed of (c + 10,000) Kilometres per second. This result, of course, would be obvious.

However, this conclusion would contradict Einstein's first principle. How can it be? Because A and B are performing identical experiments within their vehicle, and they must obtain identical results. Therefore, both A and B must find the speeds of light to be c.

It can be concluded that, no matter how large the uniform motion of the vehicle is, an observer standing in the vehicle will always measure the speed of light to be c. This revolutionary hypothesis of Einstein and its ramifications eventually transformed the landscape of physics.

This example simply shows that to come up with something dramatically new, a bold hypothesis may be needed. In addition, it also demonstrates that some principles cannot be violated. In this particular case, it is the principle of relativity that one has to follow. All these principles can guide us in finding new ideas and solving problems.

We can show another example that certain rules cannot be broken. If someone claims that he has invented a perpetual-motion machine, we need not bother to waste our time investigating its usefulness, as the claim completely violates the Laws of Thermodynamics.

At the other end of the spectrum, there exist principles that we can follow in coping with various problem situations. This is particularly important when we navigate into uncharted territories that we are not familiar with. The principles allow us to reason from the general to the particular. We will explain this kind of deductive reasoning more when we come to the chapter on Induction and Deduction.

But first, let us take a look at the experiment stage of The Scientific Method, as any hypothesis needs to be tested to see whether it actually works in reality.

Chapter 5

Experiment

In scientific discipline, an experiment is a test under controlled conditions to investigate the validity of a hypothesis. In everyday language, experiment can be interpreted as a testing of an idea. In this book, we employ both these definitions. Within the context of the first definition, we attempt to confirm whether an explanation of an observation is correct. Within the context of the second definition, we check whether a proposed idea for a solution is valid.

In scientific endeavors, an experiment is usually performed to test a hypothesis regarding how one variable (the dependent variable) changes with respect to another variable (the independent variable). In performing the experiment, care should be taken that the independent variable is the only factor that varies. In that sense, the experiment is said to be appropriately controlled.

To gain any scientific merit, the experiment needs to be reproducible, i.e., the experiment can be reproduced by someone else working independently.

In daily life, we do occasionally experiment in the scientific sense (e.g., cooking). However, most of the time, we experiment in the sense that we would like to test whether an idea would work in getting a problem solved. As long as the idea works, it may not be necessary to reproduce the experiment. We will demonstrate both types of experimentation in the following examples.

Example 1 Male impotence

In the late 1990's, a pharmaceutical company introduced the V-pill for treating erectile dysfunction, which is commonly known as male impotence. Erectile dysfunction, according to one study, increases with age and by age 45; most men have experienced it at some point in time. The V-pill should be taken about one hour before sexual activity, and it lasts for about 4 hours. Henry, in his late fifties, soon discovered that he could not do without them.

About five years later, another pharmaceutical company came up with the C-pill for impotence treatment. The C-pill can work up to 36 hours. It costs US $11.50/tablet, about 10% more than the V-pill, which costs US $10.50/tablet. However, the C-pill can last nine times as long as the V-pill. After carrying out the cost-benefit analysis, Henry switched to the C-pill.

While the C-pill did deliver what it claimed, Henry noticed that he got a headache for about 24 hours after taking the pill. He discussed this with his doctor. The doctor told him that these pills tended to increase the blood flow of the body in general and that must have caused the headache. He said that some of his patients had the same adverse effect, but they did not seem to mind. One of them told him that as long as he could have sex, a little headache would not discourage him. Nevertheless, Henry found the headache quite uncomfortable.

About two weeks later, Henry was having lunch with Tom, a close friend of his. He told Tom about his experience with the headache. Tom listened, and then proposed that Henry only took half a tablet. Henry thought that was a good idea, and wondered why he didn't ever think of it. Following Tom's suggestion, he tried taking only half of the pill, and found that it could work up to 24 hours, meaning that he could perform two days in a row. Not only did this simple idea of Tom eliminate his headache, it also saved him about $50 a month.

Later on, he experimented with taking only one quarter of a tablet, and observed that it could work up to about 4 hours. He also experimented with taking one-third of the tablet, and tried to find out how long it would last, and what was the exact time for its optimal performance.

Henry discovered subsequently that, the pharmaceutical company that introduced the C-pill did come up with tablets having only ½ and ¼ of the original dosage a couple of years later.

Example 2 Cooking

Cooking provides lot of opportunities for experimentation. As a matter of fact, cooking a dish draws a lot of parallels with performing an experiment.

In a scientific experiment, physical objects, chemical compounds or biological species are chosen for study. Samples are then prepared. Apparatus, which can be chemical, biological, mechanical, electrical, magnetic or optical, is then applied on the sample by following certain experimental procedure. Certain parameters, e.g., electrical field, can be varied to produce various end results.

Similarly, in cooking a dish, the ingredients, e.g., meat, are chosen. They are then cut and marinated. A cooking utensil, e.g., pot, pan, or wok will then be employed to cook the food by following a certain recipe. Variations in spices added, cooking temperature and time duration, would produce different end results in the taste of the dish.

For meat dishes, one of the end results would be how well the meat will be cooked. For beef, we can have rare, medium, and well done. However, for pork, chicken, and especially seafood, they should be cooked just right to taste mild, moist, and tender. If overcooked, they taste dry, tough, and stringy, and if undercooked, they risk being a health hazard as they may be infested with bacteria.

Cooking seafood, especially fish, is particularly tricky. Cooking time, quite often, needs to be controlled to within half a minute so that it will not be overcooked. The fish can be cut into filets or even smaller pieces, and the chef can bake or stir fry them, and sample taste a piece to see whether it is cooked or not. However, for a whole fish, whose size, shape and weight vary from one to the other, the chef would usually finds it difficult to cook it just right, as he, quite often does not know how long he should cook it to make it just tender. For a whole steamed fish, chefs in restaurants tend to overcook it so that the customers would not complain that it is undercooked and request to have it taken back to the kitchen. According to the experience of a gourmet, 95% of the steamed fish that he has eaten in restaurants are overcooked. That is why, if one wants to eat a not-overcooked steamed fish, one has to prepare it at home, and experiment with the temperature and time to get it cooked just right.

Charles lives in Canada. Every year, he flies back to Hong Kong to visit his mom. Just like his mom, he loves to eat steamed fish. His mom's maid, Rosini, cooks excellent steamed fish. Every day, she purchases the same kind of fish, weighing 1¼ lb (approx. 0.57 Kg), uses the same cooking utensil on a gas range, turns the heat control to the same position, and steams the fish to exactly six minutes. When Charles was in Hong Kong, he would eat steamed fish with his mom just about every day.

Back in Canada, he would like to repeat what the maid did. However, he had only an electric range. An electric range cannot produce such high heat as a gas range, nor does it have the instant on/off heat control. Nevertheless, that did not seem to concern him, as he believed that fish should be steamed in medium heat. Using medium heat, he thought, would render the timing not so critical, i.e., he could allow a larger error margin in time than if he had used high heat. His experiment showed that it would take about nine minutes to have a fish, weighing 1½ lb (approx. 0.68 Kg), fully cooked. Unfortunately, for the several times that he tried, the meat always tasted dry and tough. He could not understand why. He did try reducing the cooking time, but then the fish would be slightly raw in the centre, and thus, not fully cooked.

Nonetheless, he did observe that, every time after the fish had been steamed, there were about 50 cubic centimetres of water in the dish. He presumed that the water came from the steam condensing back onto the dish, and only part of the steam went through the steam escape hole on the lid. However, in order to prove that his explanation was correct, he repeated the whole steaming procedure without a fish on the dish. To his surprise, after nine minutes of steaming, there was only about 1 cubic centimetre of water in the dish. This simply meant that almost all the steam had come out through the steam escape hole on the lid. Thus, the water in the dish of the steamed fish must come out from the fish itself. That means, the more the water comes out, the drier would the fish taste.

Charles later found out why that happened. Heat shrinks muscle fibers, and causes water to squeeze out of the fibers, thus making the meat dry and tough. More than half of the water is squeezed out of the meat between 140° F (60° C) and 160° F (71° C). Therefore, the trick to have tender cooked meat is to turn the heat on as high as possible, so that the meat is fully cooked as quickly as possible, such that as little water as possible comes out of the meat.

From then on, Charles turned the heat control of his electric range to very high, and cooked the fish for eight minutes. As a result, the steamed fish turned out to taste much tenderer.

The same high temperature principle would apply to broiling and barbecuing; both cooking techniques can produce meat that is softer and more succulent than baking in the oven. The baking temperature in the oven is usually set to about 350° F while broiling temperature can go up to about 550° F and barbecuing temperature can go up to above 600° F. However, as the heat from both broiling and barbecuing comes from one side only, the pieces of meat have to be flipped at approximately half of the total cooking time to get the other side of the meat cooked as well. In addition, it would be convenient to prepare the pieces of meat to have roughly the same thickness so as to assure that they are all fully cooked at the same time.

Stir frying also uses the same high temperature principle. The meat is cut into small pieces — bite size. A wok, which is a round-bottom iron pan, is heated to a temperature of about 400° F. Some oil is then poured down the side of the wok. Dry seasonings, e.g., ginger, garlic, etc. are then added. When the seasonings can be smelled coming out of the wok, meats are poured in and stirred. As stir-frying uses high heat, the pieces of meat must be large enough to cook through without burning, but small enough that it will take only a few minutes to cook such that a least amount of water will come out of the meat. Since the stir-frying takes only a short time, flavor and texture are retained.

Example 3 Eye floaters

About five years ago, David woke up one morning and noticed that there were floaters in his eyes. Floaters are small spots or clouds moving in one's field of vision. They are actually small lumps of gel inside the jelly-like fluid that fills the inside of our eyes. This obstruction of vision bothered him a lot, especially when he was reading.

How did this happen, he wondered? How did he get floaters all of a sudden? What had occurred to him, or what had he done in the last month or so?

It was November, and it was about two weeks after Halloween. Halloween is a custom celebrated on the night of October 31. Children would dress in costumes and go door-to-door to collect candies. Stores would sell bags of candies to people who would like to distribute them to the kids.

After Halloween, stores would put the leftover bags of candies on sale. One store was selling chocolate candies for a discount of 50%. As David was a chocolate-lover, he bought packs and packs of chocolate, and had been eating about 50 grams per day for the last ten days or so. Could it be the chocolate that was causing the floaters?

To test his hypothesis, he stopped eating chocolate. The floaters went away in a few days. By coincidence, David had an eye checkup by his ophthalmologist in about two weeks. He told his ophthalmologist what had happened. However, he was assured that diet had nothing to do with floaters. Nevertheless, David was not going to go back eating lots of chocolate, as he did not want to take a chance. He limited his consumption of chocolate from then on and would only eat about 20 gm once in a while. To his relief, he had not seen any floaters ever since.

In the true spirit of scientific investigations, David should go back to eating lots of chocolates, in order to verify whether the floaters would come back. He should repeat the experiment several times, to ensure that it was the chocolate and not other events that might have caused the floaters. He could also vary the amount of chocolate he ate and see what amount would trigger the recurrence of floaters. This, of course, is what he should do if circumstances allow him to do so, and if it is not very inconvenient for him to do such an experiment.

However, in real life, this kind of experimentation probably rarely happens. Quite often, we would check out whether a hypothesis works. If it does, then we consider the problem solved. If it does not, then, we would search for another hypothesis and thereupon set up a test for it.

Nonetheless, we should be careful that sometimes it could be a coincidence that a certain idea actually solves a problem, as there may not be any correlation at all. There is this story that a cock wakes up in the morning and starts crowing. Every time he crows, the sun rises. Consequently, he is so proud of himself that he can cause the sun to rise, that he makes sure that he wakes up every morning on time to crow.

Thus, a hypothesis should be well tested before we accept that there is certain validity in it. If a hypothesis that we believe in does not explain any future phenomenon, do not be stubborn and attempt to defend it. Instead, try to modify the current hypothesis, or start afresh and look for another explanation.

Example 4 Restaurant dinnertimes

Lucy lives in London, England. Every other year, she flies to Hong Kong (HK) to visit her younger brother, Johnny. Brother and sister, together with other family members, would eat out quite often.

Johnny is a food connoisseur. He has a discriminating taste of what he eats. He collects newspaper articles of food critic columns. If a restaurant is recommended by a food critic, he would not mind spending HK $200 taxi fare to go all the way out to a small restaurant to eat a meal that costs HK $40. Nor would he mind spending HK $18,000 (approx. US $2,300) for dinner for a table of eighteen people, as long as he thinks the food is excellent for his taste.

There are quite a number of good restaurants in Hong Kong, as Hong Kong people are quite picky in what they eat. At dinnertime, one can choose from a regular menu listing, or a set banquet menu. Banquets are usually ordered on special occasions, like birthdays and weddings. The dishes are planned by the master chef so that each course has a different favor, thus making the combination of dishes a well-balanced feast. The banquet concept has become so popular that people would just order a banquet feast for whatever reasons, as long as there are enough people to eat all the food. A banquet usually contains more than ten courses. It quite often starts off with hors d'oeuvres, which consists of cold cuts of meat and vegetables. Then come various entrees, which would include scallops, shrimps, shark fin soup, chicken, duck and fish. The banquet would then end with noodles, fried rice, dessert, and fruits. Each course is supposed to be eaten separately, i.e., the next course will be served only after the customers have finished eating the present course on the table. In some of the high-end restaurants, the plate of each customer will be changed for each course, so that the food favors of the courses will not mix.

A restaurant's reputation quite often depends on what kind of banquets they can put forward. In Hong Kong, there are quite a number of reasonably priced restaurants that serve good banquets. As such, they are very popular and very busy at dinnertime. One would not be able to reserve a table any time that one wants to. The restaurants require the customers to reserve tables either at 6 pm or 8 pm, such that they can serve two rounds of customers, and make the most profit. Most customers choose 6 pm, as 8 pm is considered rather late for dinner, especially as one may have to wait for fifteen minutes before food is actually served. The customers would be asked to come on time, as the restaurants may not reserve them a table if they are late.

Like most customers, Johnny would reserve a table at 6 pm, and arrive on time. For a table of ten to twelve people, he would order from a set banquet menu. A few minutes after the order had been in, the d'oeuvres were served. Then the courses kept coming, at a rate of about two or three minutes apart. That, of course, did not allow the customers to finish one course at a time, which was what most customers would prefer. This just happened every time that Lucy went out for dinner with Johnny. On some occasions, Johnny complained, and the waiter simply took some of the dishes back into the kitchen, and had them wait on the kitchen counter-top. That, of course, was not what Johnny wanted, as food tasted best right after it was cooked, and not after it had sat for some time. A couple of times, Johnny instructed the waiter before he ordered to arrange each course to come at approximately ten minutes apart. However, the order was ignored. And Johnny wound up pretty annoyed.

Lucy then told Johnny that there was no point in getting upset. And why couldn't he do something about the situation? Johnny simply replied that there was nothing he could do, as the restaurant kitchen controlled when the dishes were cooked.

However, Lucy got an idea, and she wanted to test it out. So she invited Johnny and other family members out for dinner, and she planned to order a banquet. She had a hypothesis. She reasoned that kitchen staff was always busy, especially right after 6 pm, when a lot of orders came in. The staff would try to clear as many orders as possible when they came, not knowing how many orders would come in later. That was the reason why the staff was rushing the dishes out. Nevertheless, she had an idea. She arranged everyone to arrive at 6 pm, and they sat around and made small talk. She put the order in at 6:30 pm, figuring that the kitchen staff was extremely busy at that time, and would not be able to rush any order out even if they wanted to.

The first course came at about ten minutes later, and the following courses came at about ten minutes apart until about 7:50 pm, when the last few courses came together. But by then, everybody was just about full, and it did not matter that the last few courses came at the same time. They eventually finished dinner at 8:30 pm and left. In principle, the customers are supposed to finish before 8 pm, and allow the next round of customers to start at 8 pm. However, in practice, as most people do not like to have dinner at 8 pm, restaurants are usually not full after 8 pm, and waiters would not kick the 6 pm customers out. In any case, even if they really had to give up the table for the next round of customers, they still could have done so.

Lucy's philosophy was that there was no point in keep complaining about a situation. One should come up with a hypothesis of what actually is happening, and then act on it. That does not mean that one will be right, especially in an unfamiliar environment, but that is better than not doing anything at all.

Example 5 Trading the financial market

Traders monitor the financial market by using technical indicators to forecast how the market will be doing. They employ certain trading tactic to get in and out of the market.

In the year 1998, Catherine got a call out of the blue from an investment manager. He told her that he had devised a trading methodology to trade the S&P (Standard and Poor) futures. He had backtested the data for the last twenty years, and had found that his technique was quite profitable. All he needed to do now was to religiously follow his optimized technical indicators. Would Catherine like to invest in the fund that he would be starting?

Catherine asked the manager to call her back in three months, as she would like to see how his fund would be doing. Five months later, the manager called and told her that he just made a huge gain the day before. Catherine asked him how much had he gained since inception. There was then dead silence on the other end.

Catherine then asked him to call back when and if he had made a positive gain in his fund since inception. The manager never called back. Presumably, his experiment on his idea had failed.

Example 6 Silent auction

Bob is a member of a fitness club that is one of the largest fitness chains in Canada. The membership fee is reasonable and, in order to avoid having the member paid out a large sum of money for the annual membership fee, the club arranges to automatically withdraw from the member's bank account on a bi-weekly basis.

The Ottawa branch that he frequents usually holds an annual fund-raising event every year in February, with the proceeds donated to a certain charity organization. In the year 2007, the fund-raising event was chosen to be a silent auction with a starting date of February 12 (Monday) and a finishing date of February 18 (Sunday). Some of the items being auctioned were training lessons with several personal trainers. One item that stood out was a one-year membership which would worth about $800. The starting bid was $300. In order to bid on the item, the bidder would write his/her name, phone number and the bid price on the corresponding bidding form that was laid on a table. All bids were available for others to see.

Bob was interested in bidding on the one-year membership. So, he double-checked with one of the staff about the exact time that the bidding closed, and was told that it would be 6 pm on February 18 (Sunday). 6 pm is the closing time of the branch on Sundays.

As the bidding had a definitive close time, it would be obvious who the winner of the bid would be — it would be the last bidder that bid before the scheduled close time. At least, that was what Bob thought.

So Bob arranged to arrive at the fitness club at 4:30 pm on February 18, and exercised for about an hour. At 5:45 pm, he put in a bid of $500 on the one-year membership auction form, noting that the previous bid was only $490. He then went to the change room to change. When he came out of the change room, the clock on the wall read 6:01 pm, and the assistant manager was the only person manning the reception area at that time. The reception area was facing the auction table. He checked the auction form, and indeed he was the last person who bid on the one-year membership. In theory, he should win.

For the next week, he waited for the club to contact him. But no one did. So he went to the club. There, he ran into the assistant manager and he asked her what the highest bid for the one-year membership was. She told him that the bidding had been closed, and the close time was 6 pm on February 18. He repeated whether the close time was 6 pm, and she confirmed that was indeed correct. Bob then asked her again what the highest bid was, and was told it was $510. He requested to take a look at the bidding form. She hesitated, but eventually showed him the form, after quickly covering the name of the last bidder.

Bob then told her that at 1 minute past 6 pm on February 18, he checked the bidding form of the one-year membership, the last and highest bid was $500. The assistant manager then told him, somewhat to his surprise, that while the club was officially closed at 6 pm, some members might still be in the club, and they would be allowed to bid if they so wished. $510 was the highest bid and that was final.

Bob thought that the whole process was somewhat dubious. Noticeably, why would the assistant manager cover up the name of the last bidder, as this was an open bid and all members should be able to see the names of all bidders? So he wrote an email to the club's Headquarters in London, Ontario, urging them to look into the matter. Headquarters forwarded the email to the manager at the Ottawa branch. The manager replied the next day, simply saying that the assistant manager had made a mistake. The close time of the silent auction was not 6 pm on February 18, but 12 pm (midnight) on February 19 (Monday). (12 pm is the closing time of the branch on Mondays.)

Bob thought that argument was ludicrous. So he wrote an email back to the manager, with a copy sent to the Headquarters. He reminded her that the closing date February 18 was clearly written on the information board beside the auction table, and not February 19 as she claimed after the fact. That was indeed the case had been verified by other members. Furthermore, the assistant manager repeated twice that the close time for the auction was 6 pm on February 18. As a matter of fact, she also said that the bidding forms were collected early Monday (February 19) morning.

That email was never replied, not by the manager, nor Headquarters. A few days later, Bob asked the manager whether he could take a look at the bidding form again, and was told that it had been discarded. Bob eventually decided not to pursue the matter any further as he had more important things to do.

This example just shows that an idea is just an idea, and may not work out in reality. One may think that the last highest bid of an auction before close time would be the winning bid. But that may not be the case if the bidding form is tampered, and authority turns a blind eye to such an action.

Quite often, we may come up with an idea and believe that there is no particular reason why it would not work. But the sad fact is, there can be many obstructing factors that we are not aware of, or that are completely out of our control. A professor once told his new graduate student that "Many ideas look good only on paper". The student later found out that it was indeed very true. He quite often woke up in the morning filled with ideas, only to find out that most ideas would not work out when he tested them.

Nevertheless, though many ideas do not work, it is the ones that do that make a difference. Even if only 10% of the ideas work, it is still much better than not having any idea at all.

5.1 Experiment versus hypothesis

It should be noted that in scientific investigation, or everyday problem solving, hypothesis does not necessarily have to come before experiment. It can come after as the experiment may need to be performed first, and observation done before a hypothesis can be drawn. Now the question is, at what point should one come up with a hypothesis? Should one attempt to collect lots of data, analyze them before proposing a hypothesis, or should one jump into presenting a hypothesis before even getting any experimental data? We believe that one should suggest a hypothesis with the minimum possible amount of data. The idea is that we should attempt to find an explanation or solution with the minimum amount of time and resource, i.e., we would try to get to the goal as quickly as we possibly can, and with as little effort as we can dispense.

An example of quickly coming up with a hypothesis in scientific research is the discovery of the basic structure of DNA in 1953. DNA had been a mystery, and it was up for grabs for anyone who would like to give it a try. At the University of London, Maurice Wilkins and Rosalind Frankin were busy taking X-ray diffraction photographs of the DNA molecule. They figured that they could build a model of its structure after they had collected more experimental data. In the mean time, James Watson and Francis Crick at the Cambridge University believed that there might be enough data already. The structure could be discovered by a combination of guesswork and child-like model building. After some trial and error, they produced the double helix, which was the answer to the structure of DNA. Finding the structure of the DNA is considered to be one of the significant discoveries in the twentieth century.

Hypothesizing is an active process that requires the diligent use of our brain. It forces us to think and come up with an explanation or a solution. The prediction from a hypothesis would direct us to more observation and experimentation so as to confirm whether it is correct or not. Even if it is incorrect, it is still useful as it allows us to eliminate it as a possibility, and steers us to search for other avenues. Hypothesizing serves as a guide to our final destination. However, a hypothesis requires careful and meticulous experimentation to verify that it is sound. In the next section, we will take a look at the history of the development of the balance between hypothesis and experiment.

5.2 Platonic, Aristotelian, Baconian, and Galilean methodology

Plato (427–347 BC), building upon the teachings of his teacher, Socrates, argued that reality was eternal and unchanging, and could be found only through reasoning in the human mind, and not through our sensory experience. As a matter of fact, he believed that our sense impressions could deceive us. He was convinced that we were born with knowledge, all we needed to do to come up with the truth was to sit, think, and discuss with others.

Unlike Plato, Aristotle (384 BC–322 BC) believed in empiricism — that knowledge came from one's sensory experiences. He endeavored to hypothesize at an early stage of scientific investigation, but unfortunately, he did not attempt to confirm his hypothesis with further observations (e.g., he said erroneously that women had fewer teeth than men). Furthermore, he did not try to prove his hypothesis by performing experiment (e.g., he incorrectly claimed that heavy objects fell faster than light objects).

While we can arrive at some of the truths by observing nature with our eyes wide open, most of the truths will not come by unless we purposely design our surroundings to entice them out, i.e., we perform experiments. Francis Bacon (1561–1626) is one philosopher that emphasized experimentation. He proposed that truth should only be derived from careful collection and interpretation of data after performing detailed experiment. While his method will lead to a very systematic accumulation of information, it underplays the early proposal of hypothesis.

Galileo (1564–1642) established the practice of quantitative experiments, and analyzed the results mathematically. His experimental method is what most scientists would associate with in the modern sense. He would use an experiment to prove whether a hypothesis was correct, needed correction, or simply should be discarded.

In summary, Plato hypothesized but did not observe. Aristotle hypothesized after some observation, but did not pursue further observation, nor experiment to confirm his hypotheses. Bacon proposed detailed experimentation, and suggested hypothesizing only when one was certain that one's copious experimental data would support one's hypothesis. Galileo hypothesized, and performed experiment to verify that his hypothesis was correct.

In conclusion, we believe that, unlike Bacon, we should come up with a hypothesis as quickly as we possibly can, and, unlike Aristotle, we should carefully and meticulously experiment to confirm that it is right.

Now, before we can hypothesize and experiment, we need to recognize that a problem exists to begin with. We will study problem recognition in the next chapter.

Chapter 6

Recognition

Before we can solve any problem, we need to recognize that a problem exists in the first place. That may seem obvious, but while some problems stick out like thorns in a bush, others are hidden like plants in a forest. As such, not only do we need to tune up our observational skills to see that a problem does exist; we should also sharpen our thinking to anticipate that a problem may arise.

Example 1 Electricity blackout

It was Christmas of the year 1998. A Canadian family of four went to the United States for a vacation. As they were heading back to Canada, they passed through New York State, and checked in a motel in a small town.

As it happened, there was an ice storm in the area. In the middle of the night, there was an electricity power blackout. The family woke up in the morning, and realized they had lost electricity. They then decided to have a quick breakfast, and get back on the road as early as possible. They had a ten-hour drive ahead of them, and they preferred not to drive at night. As the motel served complimentary Continental breakfast, they chose to head down to the lounge where breakfast was served. (A Continental breakfast is a light breakfast usually consists of croissants, bread, pastry, coffee, and tea.)

As they were heading out of their motel room door, their twelve-year-old son remarked that they might not be able to get back into the room, as the electricity was out. The motel was using plastic door-key cards with magnetic stripes. The father then suggested that he would go outside the room by himself, and try to open the door with the key card and see whether the door would open. He soon found out that it would not. Had all four of them gone out of the motel room, they would not be able to come back inside for their luggage.

They thought of taking their luggage to the car first, and then going for breakfast. However, they would prefer to brush their teeth after breakfast, and someone might want to use the bathroom also. Eventually, the mother and daughter decided to stay in the room so that they could open the door from the inside. The father and son then went down to the lounge to get some croissants and bread, and took them back to the motel room. They had breakfast and then checked out of the motel.

This is a good example of why it is important to anticipate potential problems. Had the son not recognized the relation of the magnetic key card with the power outrage, and alert to a possible inconvenient situation, the family might have stuck in the motel for a few hours.

Example 2 Car skidding

The Smiths lived in Winnipeg, where the temperature in the winter could go down to –40° C, and the roads could be icy and slippery after it snows. They owned two vehicles. Most of the time, the wife, Nancy, drove the car, and the husband, Charles, drove the van.

One winter evening, Nancy came home from work and started yelling at Charles, "I told you to look after my car, but you did not. It skidded again this morning, and this is the third time that it has skidded this winter." This was actually the first time that Charles had heard from Nancy that her car had skidded. She then told him that the car skidded in the same spot every time. That morning, it skidded, banged onto the curb, and turned a full 360°. Fortunately, no one was hurt.

The Smiths lived in a neighbourhood where there was a winding four-lane road joining a main road that led to the city downtown, with two southbound lanes coming in and two northbound lanes going out. One side street intersected the southbound lanes. Unfortunately, there was a curve about 20 metres north of the intersection where the side street cuts the main road. For cars coming out of the side street, and wanting to head north, the curve made it difficult for them to see the southbound traffic. Once the drivers saw that the traffic was clear, they had to accelerate quickly, and then slow down abruptly at the left northbound lane in order to turn north. The spot where they had to slow down was the exact spot that Nancy skidded on.

Why that particular spot was slippery was obvious to Charles. Increasing the pressure on ice or snow lowers its freezing point, and turns the ice or snow into water. (This is actually the reason why people can skate on ice.) When the cars decelerated abruptly, the tires put pressure on the snow covering the road, melting it into water, which then froze back to ice, thus making that spot slippery. Charles had previously cautioned Nancy to slow down before stop signs, as those spots were particularly slippery. However, Nancy failed to see the similarity between spots before the stop signs, and spots where cars had to decelerate quickly. She should have slowed down at the intersection if she was driving on the left northbound lane, as she would hit that particular slippery area. Better still, she should drive on the right northbound lane when she approached that intersection.

In any case, Charles wondered why Nancy had to wait till the third time she skidded before raising the alarm. It seemed as if Nancy never realized that there was a problem to begin with.

Not recognizing a problem, or not recognizing how serious a problem is, can be serious, as the following two examples will show.

Example 3 Eye vision

James was in his early eighties and in good health. Nevertheless, he began to have problem doing close work like reading. He had noticed that the central vision of his right eye started to blur, and he just assumed that it was a normal part of aging. In any case, his left eye could see very well. A year later, his right eye was getting worse. It still had good side vision, but blank spots appeared in the centre. One day, his daughter was visiting him from out of town. He mentioned his problem to her, and she urged him to go see an ophthalmologist.

As it happened, he had an age-related macular degeneration (AMD). Macular degeneration is the damage of the macula. Macula is the part of the retina that is responsible for the sharp, central vision needed for reading and driving. The retina is a light-sensitive membrane lining the inner eyeball, and is connected by optic nerves to the brain. Unfortunately for James, his AMD was in an advanced stage, and there was not much the doctor could do.

There was a possibility that he might be blind in his right eye in a few years.

James should have arranged an eye examination by an ophthalmologist every two years or so. As the saying goes, prevention is better than cure. Unfortunately, he did not recognize the seriousness of the problem until it was too late.

Example 4 Flu?

David was an engineer in his late thirties, and his wife was a nurse working part time in a hospital. One day, he had a fever, and thought he might have the flu. He went to see his doctor, who confirmed that David had the flu. That night, his temperature shot up to 104° F, but in the morning, it had come back down to below 100° F. So, he thought that he was getting better. Unfortunately, the high temperature the night before was a warning signal that he missed.

In any case, he called in sick from work. As the day went by, his wife noticed that he was getting disoriented, and could not even recognize their children. After consulting with a nurse friend of hers, she decided to drive David to the emergency ward of the hospital she worked in. As David was waiting in a bed at the hallway, his wife ran into one of the cardiologists, Dr. Jones, whom she worked with. She asked Dr. Jones whether he could take a quick look at her husband. Dr. Jones tried to talk to David, but he did not respond. And then suddenly, David went into respiratory arrest. It was fortunate that he was in the emergency ward, and Dr. Jones was right beside him. He was immediately hooked up to a machine to keep him breathing, as, without oxygen to the brain, one could easily become a living vegetable in 5 to 10 minutes.

David then began to have seizures, and eventually went into a coma.

He was quickly taken to the intensive care unit. Several specialists gathered together to figure out why he was in a coma, and what was wrong with his brain. They took a CAT (computerized axial tomography) scan of his brain to check whether there were any blood vessels that had burst. They also did some tests on his spinal cord fluid to check for bacteria that might cause meningitis. Both results were negative. They eventually treated the case as encephalitis, which was an acute inflammation of the brain commonly caused by a viral infection. They did check for virus, but could not find any. However, not finding any did not mean it was not there, as it was not at all easy to detect a virus. In any case, the doctors kept him breathing with a tube through his throat. They gave him anti-virus drugs, and lots of IV (intravenous) fluid.

He woke up after three days, and thought that he had a serious car accident. He lost his short-term memory, which, fortunately, mostly came back later. Further test with MRI (magnetic resonance imaging) did not show any damage to his brain.

David was lucky. His wife was a nurse, and was with him when he got disoriented. She recognized the seriousness of the problem, and quickly took action. If she had not, he might got to the point of no return, and became brain dead.

Sometimes, we may recognize a problem, but do not realize that there is a solution, or a solution can easily be found. Thus, it is also important that we should recognize that a solution might exist, and attempt to look for it.

Example 5 Central heating

Greg and Liz live in a two-story house in Hamilton, Canada. Most of the houses in Canada come with central heating, where hot air is pumped through a system of air ducts into all the rooms from a furnace located in the basement. The desired temperature is controlled with a thermostat, which will automatically turn the heating on and off to maintain the rooms at the set temperature.

One weekend morning in the winter, Liz told Greg that her throat always felt very dry in the morning, and it had been like that for the last one month. She figured that it had something to do with the hot air blowing from the furnace all night. Greg thought that she might be right, as the hot air might not be humid enough. Humidity is the amount of water vapour in the air. It is needed for the comfort and health of the people living in the house. Too little humidity can cause chapped skin and dry throat.

There was a central humidifier mounted on the cold air duct connected to the furnace. Water was fed automatically to the humidifier, making the hot air humid. Although the humidity control could be manually adjusted, the adjustment was always a bit tricky. If the humidity was too low, the hot air would be too dry. If the humidity was too high, the excessive humidity could cause damage on walls, ceilings and floors, as well as forming mildew on their surfaces. Greg usually tried to adjust the control slightly on the low side to avoid any possible damages.

Now it seemed that Liz recognized that there was a problem, but she did not realize that a solution could be easily found, and therefore she had not mentioned any displeasure for the last month. Greg never knew Liz had a problem. But now that she mentioned it, he thought that the problem could easily be solved.

They usually set the thermostat temperature to 18° C all night. Thus, one easy solution was to set the thermostat higher to 21° C an hour before they went to bed, so that the house was warmed to 21° C, and then set the thermostat back to 18° C just before they went to bed. That way, there would be much less hot air blowing through out the night. Alternatively, they could simply set the thermostat temperature from 18° C to 15° C before they went to bed. That would perform the same function, as well as reducing the gas bill. Or better still, they could simply shut off the heating system before they went to bed, so that no hot air would blow through. This last suggestion might sound a bit drastic, but it all depended upon one's point of view.

People living in North America have been quite pampered. They have central heating. In many parts of the world, not only is there no central heating, there is basically no heating at all.

Greg has his share of experience with heating, or rather, no heating, in houses in other parts of the world. When he was a kid living in Hong Kong in the 1950's, his family was definitely not well off. They lived in a small flat, with a veranda (a balcony) that was open to fresh (or rather, polluted) air. In winter, the temperature could drop to about 3° C. There was no heating in the house. To make it worse, since there were ten of them in the family, there was not enough space for everybody to sleep inside the flat at night, and he had to sleep outside in the veranda. He remembered he had to sleep with his socks on, as well as wearing his woolen undershirt, and cotton jacket when it got very cold in the winter.

But he considered himself lucky. A classmate of his, Chan, had to sleep on a piece of wooden board placed on a water tank. Back then, the water supply in Hong Kong came from rainwater stored in reservoirs. If it did not rain for a few months, water had to be rationed. So each family would keep a water tank in their own rented room, where a family of six or more would live in.

One night, while Chan was sleeping, he somehow fell into the water tank, and got himself all soaked.

Greg later went to Canada to study at a university. And then much later, in December 1979, he was invited to work as a research fellow at a university in Britain. So, he flew to the Gatwick airport in London, and then took a train to Brighton. After Greg stayed in bed and breakfast in Brighton (which is close to the University) for a week, he found himself a rented room. However, every morning at around 4 am, he was woken up by the coldness of the room. During a coffee break at work, he kind of mentioned that his landlady did not turn the heating on at night. One of the postgraduate students then told him that was the norm in Britain. Natural gas was expensive, and the British people only turned on the heating for a few hours after they came home from work, and turned it off before they went to bed. Greg had been living in Canada too long to realize how fortunate he had been.

Now that Liz had complained about a dry throat in the morning, Greg suggested that might be they should set the temperature from 18° C to 15° C before they went to bed. In that case, the heating would not be automatically turned on by the thermostat as often as if they had set it to 18° C all night. After they made this adjustment, Liz no longer woke up with a dry throat. Furthermore, the lower temperature in the house would also contribute to reducing global warming pollution.

Thus, not only should we recognize that a problem exists, we need to recognize that it is possible that a solution can be found. Quite often, we may have to rephrase the problem in a way that a solution can be attainable.

In the following, we will see how a twenty-year old non-economist spotted a problem while sitting in an undergraduate economics course. He recognized that there could be a solution, and phrased the problem in a manner that it could be solved. He worked out the solution and submitted two papers for publication 1½ years later. Those journal papers eventually netted him an Economics Nobel Prize.

6.1 **John Nash** (1928–)

John Nash is undoubtedly one of the mathematical geniuses of all times. Originally he wanted to follow his father's footsteps to be an electrical engineer, but eventually decided to major in chemical engineering when he attended Carnegie Institute of Technology in Pittsburgh, Pennsylvania in 1945. He soon found that chemistry was not particularly interesting, and, with the encouragement of the mathematics faculty, switched to majoring in mathematics. He was able to make so much progress in his mathematic courses that he received both his bachelor's and master's degree in 1948.

When he applied to Princeton for graduate school, his Carnegie Technology professor, Richard Duffin, wrote only five words in his letter of recommendation: "This man is a genius".

While at Carnegie, he took an economics course in his final semester, simply to fulfill degree requirements. That was the only economics course he would ever attend. While sitting in that course, he recognized an unsolved bargaining problem concerning trade deals between countries with separate currencies. After he arrived at Princeton later, he would work out the details.

Princeton in 1948 was a heaven for mathematicians. Its Institute for Advanced Study was where the bright stars, Einstein, Gödel, Oppenheimer, and von Neumann did their mind-boggling work.

Von Neumann (1903–1957) is a genius in his own rights. In the 1920's, he invented Game Theory. The objective was to build a mathematical theory of rational human behavior using simple games as examples. Together with Oskar Morgenstern (1902–1977), he wrote the book *The Theory of Game and Economic Behavior*, which was published in 1944. The book was considered to be "the bible" by economics students.

The book comprised of methods for finding mutually consistent solutions for zero-sum two-person games. The focus was on cooperative games, assuming that the players could come to an agreement among themselves with optimal strategies. However, to Nash, the book contained no basic new theorems other than Neumann's min-max theorem, which guarantees that each player of a zero-sum game has an optimal tactics.

Nash figured out a way to generalize the min-max theorem. It did not have to be a zero-sum game, nor did it have to involve only two people. He was able to provide a simple and elegant proof of a non-cooperative equilibrium for a game of multiple players. The solution would be stable, and it would not pay for any single person to deviate from the equilibrium strategy. This is because individual self-interest can override the common good, leading to a worse outcome for the whole group. Thus, participants should strive to find the information that would lead to a deal that makes rational sense for everybody involved. The consequence of this game theory extension later echoed in the real world for dilemmas like overfishing, arms race and global warming.

In the fall of 1949, Nash arranged to have a meeting with Professor von Neumann to discuss his equilibrium ideas. Before he could finish a few of his sentences, von Neumann interrupted and said abruptly that the concept was trivial.

Undaunted by the disastrous meeting with von Neumann, Nash submitted a paper "Equilibrium points in N-person games" to the proceedings of the National Academy of Sciences, and then another paper "The Bargaining Problem" to Econometrica. Both papers were published in 1950.

The papers eventually formed the basis of Nash's twenty-seven-page doctoral thesis. The thesis described the definition and properties of what would later be called the Nash equilibrium. No one, not his supervisor, nor young Nash, would have thought that the thesis was going to be Nobel material.

In the summer of 1951, Nash went to the Massachusetts Institute of Technology (MIT) as an instructor in the Mathematics Department. While at MIT, he made several breakthroughs and solved some of the classical unsolved problems related to differential geometry and partial differential equations. As a result, he was offered tenure in the Mathematics Department in January 1959.

Unfortunately, at this turning point of his career, Nash was diagnosed with paranoid schizophrenia. He had to resign from his position at MIT, and was later admitted to a mental hospital. The disease practically incapacitated him for the next two decades or so.

In July 1959, he traveled to Europe and attempted to gain status as a refugee. He returned to Princeton in 1960, and, in campus legend, became the "Phantom of Fine Hall" (Fine Hall is Princeton's Mathematics Library). He would scribble strange equations on blackboards and wander on campus as a ghostly figure.

He was in and out of mental hospitals until 1970. Miraculously, he slowly recovered, and was able to do some serious mathematics again.

In the meantime, the "Nash equilibrium" had begun to crop up in various journals, and the concept was applied to many different disciplines, including economics, politics, biology, and business studies. In 1994, Nash was awarded the Nobel Prize in Economics, for his work in game theory while he was a Princeton graduate student, and for a problem that he spotted when he was a twenty-year-old undergraduate student.

Interestingly, the problem could have been recognized by many others, but it was not. Not even the legendary von Neumann realized the existence of the problem, nor its significance after it was pointed out to him. The mathematics was not that difficult, not when one compared with the many other intricate solved and unsolved mathematical problems. As a matter of fact, Nash considered it as his "most trivial work".

Nash was able to identify this important issue, word it in such a way that it was solvable and the solution eventually got him a Nobel Prize. For those who are interested in the money aspect — yes, the Nobel Prize does come with a financial award. In Nash's case, it was about one-third of a million US dollars (as he was sharing it with two other game theorists).

Thus, recognizing a problem can bring fame and financial reward. However, in order to be able to recognize a problem, one needs to train oneself to make keen observation, and keeps one's eyes open for opportunities when they occur.

In summary, not only do we need to recognize that a problem exists, we need to recognize how significant a problem is, if indeed it is significant at all. In addition, for the identified problem, we should phrase or define it in such a way that a solution can be sought. This, we will discuss in the next chapter.

Chapter 7

Problem Situation and Problem Definition

For just about any situation, we can look at it from different perspectives. Take the example of a piece of rock; it will look different from the eyes of a landscaper, an architect, a geologist and an artist.

Similarly, any problem situation can quite often be inspected from various angles, and therefore the problem can be defined differently. Our planetary system can be analyzed as a system revolving around the earth, or it can be analyzed as a system revolving around the sun. Of course, one form of analysis may be superior to others.

For everyday problem situations, not only can a situation be looked at from perspectives on the same level; it can be looked at from perspectives on different levels.

7.1 Perspectives on different levels

A situation can be examined from a specific viewpoint, or from a more general viewpoint.

Example 1 Starting a business

Gerry would like to purchase a restaurant franchise. However, he was having problem getting the finances, as the franchise required quite a large sum of money. What he did not seem to realize was that he did not necessarily have to purchase a franchise, as his general objective was to get rich. He could have opened a dry-cleaning business instead. Thus, one can go from the specific to the more general and from the more general back to another specific.

Example 2 Broccoli

A mother would like her kids to eat broccoli, but her kids hated it. So, she would try different ways of cooking broccoli. She even tried chopping it up into small pieces and mixing it up with other food. And she also tried to make broccoli soup, and get her kids to drink it. However, she did not necessarily have to feed them broccoli, as her general objective was to make sure that they were healthy and strong. She could have fed them other vegetables instead. Thus, again, if we analyze the problem situation in a more general sense, we may come up with an entirely different problem definition.

7.2 Perspectives on the same level

Most often, we can look at a problem situation from another perspective on the same level, and thus, define the problem differently.

Example 3 Housework

John and Nancy both had full-time jobs. They came home around 5 pm on weekdays, made dinner, played with their two children, read the newspaper, and watched TV. They left most of the housework to the weekend. On weekends, they had errands to run, and occasional dinner invitations and musical shows to attend. They quite often found that they did not have enough time to do housework.

One weekend, they sat together to discuss how husband and wife could try to schedule, and get the housework done more efficiently. Nancy then suddenly came up with an idea. The children were already ten and twelve years old. Why did they not get them involved also?

From then on, they got the children taking turns to do the laundry, and mop the floor. John and Nancy then found that they had more spare time and therefore more quality time with the children. Getting the children to share the housework also taught them responsibility and teamwork.

Thus, by looking at the problem situation from a different perspective, Nancy could define the problem differently, and thereby came up with a better solution. Had she limited herself to the perspective that only she and her husband could get the housework done, she would have narrowed herself to restricted possibilities.

Example 4 Car tires

Going back to the car skidding example in the last chapter, we will see whether we can look at the problem from a different perspective.

The wife's car skidded three times at the same spot on an icy road in the winter. While she blamed her husband for not taking care of her car, the husband wondered why she waited till the third time her car skidded before telling him. He explained to her why the car would skid at the same spot, and she should slow down when she approached there, or drive on a different lane.

A few days later, the husband was driving his wife's car, and he heard some noise. The next day, he took the car to a garage and had a mechanics checked what the problem was. He was later told by the mechanics that the car had been in a collision (The car actually banged on the curb when it skidded on the icy road the third time). The left rear wheel and bearing, as well as the left front ball joints were damaged and needed to be changed. That cost him about $800. The car mechanics also told him the reason why the car skidded. It was because the two front tires were almost completely worn out, and definitely needed changing. That cost him another $350. Had he been more observant, and took care of her wife's car, he should have changed the two front tires before the winter.

Thus, the whole problem could have been looked at from a different perspective. Prevention is better than cure. The husband could have saved himself quite some money. And more importantly, he could have avoided his wife getting into an accident.

Example 5 Water faucet

One day, Jim noticed that there was a water stain under his kitchen counter-top. He then noticed that water had been seeping through the broken seal between the water faucet assembly and the counter-top, and rotting the wood under the faucet.

As the faucet was about twenty years old, and looked rather unattractive, he wanted to replace it with a new and more fashionable one. So he went to a plumbing store to get a water faucet, valves and flexible tubing.

Before he put the new faucet in, he had to dismount the old one. He then found out he had a problem. There was a hexagonal nut under the counter-top that screwed the old faucet in place. Because the screw thread was getting a bit rusty, he was not able to unscrew the nut. It was also difficult to access the nut with any tools within the tight space under the counter-top, where several copper tubing were attached to the faucet.

Not discouraged, Jim went to a hardware store a couple of times to get new tools. But try as he might, he was not able to untighten the nut. After working at it on and off for several days without any success, he called his friend Tom for advice.

Tom came over with his tools the next day. Before he attempted to unscrew the nut, he tapped on it with a hammer and tried to loosen it. But even though he managed to put a wrench on the nut after, he was not able to untighten it. He worked on it for another ten minutes, and he had to give up. He then came up with an idea. He suggested that they could use an electric drill and titanium drill bits to cut the nut open.

Jim then started drilling a hole on one of the sides of the hexagonal nut. He increased the size of the drill bit as the hole got larger. Eventually, he was able to cut the nut open on one side, and the nut came off easily. After that, they were able to take off the old faucet and put the new one in.

In this case, Tom re-defined the problem. Instead of unscrewing the nut off, all they needed to do was to take it apart, by hook or by crook. That they succeeded in doing.

Example 6 Neck pain

Jim got a neck pain from working long hours at the computer. He went to see his family doctor, who told him this was not uncommon these days for just about all ages as people spent a lot of time on the computer. The doctor sent him back home and asked him to do some neck exercises every day.

Jim dutifully performed the neck exercises twice a day for about a month. However, he did not feel any improvement. One day, he saw an advertisement of a chiropractor close by where he lived. So he went to see the chiropractor.

Upon examination, the chiropractor believed that Jim's neck bones were pinching the nerves, and needed to be adjusted. He used his hands to manipulate the neck joints in order to restore the joint motion and function. He bent the neck to the left and then to the right. In bending to both positions, there was a popping sound. The sound is actually a small pocket of nitrogenous gas escaping from the joints. Between the bones that form a joint is a fluid that acts as a lubricant to allow for smooth motion of the joint. The fluid contains dissolved gases. Pressure will build up in a joint if it becomes tight. A chiropractic adjustment, while releasing the pressure of the joint, will also release the gases in the joint space, thus causing a popping sound. More importantly, the adjustment restores the proper biomechanical and biochemical components within the treated joint.

After an adjustment, Jim felt an immediate relief in his neck. Unfortunately, after an hour or so, his neck went back to being tight and painful. He continued seeing the chiropractor twice a week for two months. But he was not actually making too much progress. He let the chiropractor know how he felt, but the chiropractor simply told him to keep coming to his sessions.

While Jim believed in his chiropractor, he also believed that the neck should not be modeled as bones only. The neck should be considered as a model of bones and muscles. Even though the bones have been adjusted properly, the neck muscles can still be tight and need to be treated. Unfortunately, his chiropractor did not treat muscles, as Jim was told.

By chance, a friend of his told him about a physiotherapist who specialized in neck and shoulder pain. So Jim went to see her. The physiotherapist first asked him at what time did his neck hurt the most. Jim told her it was in the morning, right after he got out of bed. She told him he should switch to using a medium support polyester gel fibre pillow. The pillow would provide his neck a good support with a downy softness.

She then put him on a cervical harness attached to a traction machine that would apply 15 lbs of force on his neck for 15 minutes. She also figured out which of his neck muscles was tight and performed isometric exercises on his neck — by pressing her hand against the side of Jim's head, and asked him to bring his ear toward his shoulder. Afterwards, she performed myofascial release (a type of massage technique) on his neck and shoulder. Finally, she put couplant on his neck and performed ultrasound treatment with a 0.5 MHz ultrasonic transducer for five minutes.

Jim was treated by the physiotherapist twice a week for a month. On the eighth visit, the neck muscles suddenly loosened, and Jim felt much better after. He felt that had he kept going to see the chiropractor only, his neck muscles would have remained as tight as ever. It was therefore a good idea that he was treated by the physiotherapist as well.

For the next two months, Jim continued seeing the physiotherapist twice a week, and the chiropractor once every two weeks. His neck pain slowly disappeared to the point that he did not think that he needed to see them regularly. Now he does neck exercises often, and visits the chiropractor and the physiotherapist once in a while for maintenance purpose.

It is important to define or model a problem properly in order to get it solved. As in this particular example, the neck needs to be modeled as being built of bones and muscles, and not bones only. People tend to examine a problem from their angle of expertise. When presented with a problem, a mechanical engineer may look at it from a mechanical point of view, while an electrical engineer may look at it from an electrical point of view. However, a specific aspect of a problem may not represent the whole picture, and may not lead to the problem being solved.

Problems need to be modeled properly. The orbits of planets are definitely better described being modeled as ellipses rather than circles. Similarly, electrical properties of metals are more accurately determined when the interaction between conduction electrons and ion nuclei are taken into account, rather than considering the electrons to be completely detached from the ions and run freely in a vacuum as in the free electron model that was used earlier on.

There is also one more lesson to be learnt from this example. While you should definitely consult the professionals and the experts, you should also take charge of your own problems. If one treatment does not seem to work, look for other avenues. More generally, if one solution to a problem is not getting anywhere, seek for new ones.

Chapter 8

Induction and Deduction

Once a problem has been defined, we need to find a solution. To determine which route we can take, we will have to take a look at the knowledge that we already have in hand, and we may want to search for more information when necessary. It is therefore, much more convenient if we already have an arsenal of tools that have been stored neatly and categorized in our mind. That simply means that we should have been observing our surroundings, and preferably have come up with some general principles that can guide us in the present problem.

8.1 Induction

Induction is the process of reasoning that derives general principles from specific instances. As our observation of specific instances is always limited, we should be careful in arriving at the general principle through induction. Nevertheless, in everyday life, induction is a technique that comes in handy, and can benefit us if used wisely.

Example 1 Gas prices

Chris lives in Ottawa, and he is single. His mom lives in Toronto, and he drives to visit his mom every other weekend or so. Along the highway between Ottawa and Toronto, there are several service centres with gas stations and fast-food restaurants. Every time Chris stops at a service centre for a cup of coffee, he notices that the gas price there is always higher than the gas price in Ottawa, as well as in Toronto. This may be caused by a higher transportation cost to deliver gas to those service centres, or it may simply be a supply and demand situation, where drivers have to fill in gas when they are running low. Whatever the reason is, the gas prices at those service stations along the highway are higher than the gas prices in the towns of Ottawa and Toronto. And that is the general principle that he induces by way of observation.

The subsequent action is obvious. Every time Chris travels to Toronto, he fills up his gas tank in Ottawa and when he drives back to Ottawa, he fills it up in Toronto.

Example 2 Drugs and cosmetics

A grocery store is expanding its business. Other than selling the usual meat and vegetables, it is also selling drugs and cosmetics. Nancy notices that the price of a few of the non-prescription drugs and cosmetic creams are 10–25% lower than those of the drugstore that she usually goes to. She believes that the grocery store has a lower price because it is trying to draw customers to purchase non-grocery items in their store as well. In any case, she concludes and induces that most, if not all, of the drug and cosmetics items in the grocery store are cheaper than the drugstore she shops at.

From then on, she buys all her drugs and cosmetics at the grocery store.

Example 3 Grocery store sales

Mary goes to grocery shopping once or twice a week. Grocery stores usually deliver sale flyers to her home. Advertisements in the flyers last for one week, usually from Saturday to the next Friday.

Mary goes to Store A or Store B, depending on what sale items are advertised in their flyers. She notices that Store A does not put a limit on how much a customer can buy on its sale items, and therefore, the sale items are usually run out within the first two days. Store B regularly puts a limit of two per customer on its sale items, even though it does not say so on the flyers. However, they would usually remove the limit after approximately 5 pm on the last day of sale (The store closes at 9 pm).

Once Mary figures that out, she would shop at Store A on the first day of sale, and at Store B after 5 pm on the last day of sale, if there are any sales items she needs from either store.

Thus, by following the general principles that we arrive at, we can choose the action that we will pursue. However, we should be cautious about the prescripts that we induce, as any future observation can dispute their validity. Also, they may change with respect to time. To draw a parallel in science, some scientists and philosophers of science believe that scientific theories can never be proved, they can only be disproved. A general statement can never be proved from particular instances. On the contrary, a general statement can be disproved by one incompatible observation. An ancient European conception was that "All swans are white". A black swan then became a metaphor of something that could not exist. In 1697, a Dutch explorer became the first European to sight a black swan when he sailed into the western coast of Australia. The sighting thus refutes the truth that "All swans are white".

Induction, especially in everyday problems, can be viewed as a special form of hypothesizing. The induced principle can have a much wider range of application. As such, it would have a larger chance of being erroneous. Nevertheless, general principles do provide us with guidelines of how to act in certain situations. We should, of course, watch for contrary instances that dispute them, and modify them accordingly.

Presumably, our life experience is limited, and we only have limited time to make observations to induce the general principles. Can we draw upon other resources? Fortunately, we can. We can listen to other people's advice. And, more importantly, we can exploit the general theories written in scientific books. From the general theories, we can deduce a solution to the particular problem that we encounter.

8.2 Deduction

Deduction is a process of reasoning in which a conclusion is reached from some premises accepted earlier. If the premises are correct, then, by logic, the conclusion cannot be false, as the next example will show.

Example 4 Central air-conditioning

Stephen is an electrical engineer. He was lured from the United States to work at a high-technology company in Ottawa. Being single and well paid, he had lots of cash. After living in Ottawa for a year in a rented apartment, he bought himself a townhouse. (A townhouse is one of a row of houses attached by common sidewalls.)

A couple of months later, as he had some spare cash, he bought another townhouse as an investment and rented it out to several guys. It was summer, and the temperature outside was about 25° C. One day, he got a call from one of his tenants, who told him that he did not think that their air-conditioning unit was working well. They were not getting enough cold air in their bedrooms.

Most townhouses in Ottawa, including the one that Stephen was living in, have two stories, and a basement. However, the townhouse that he rented out has three stories, with no basement and with bedrooms on the third floor. As that townhouse is facing west, it can get pretty hot in the summer, and the previous homeowner had installed central air conditioning, which is designed to cool the whole house. The air-conditioner is located outside the house at ground level. Cold air is distributed in the house through air ducts into each room. It was possible that the air-conditioner had problem moving the cold air up to the third floor.

The townhouse that Stephen lived in is facing south. As it does not get very hot in the summer, there is no central air-conditioning. As a matter of fact, he had been living in apartments and houses with no air-conditioning all his life, and had no idea how central air-conditioning ran. However, he figured that the air-conditioner at the rental townhouse must be old, and either needed a tune-up or should be completely replaced. Not knowing exactly what to suggest, he told his tenant that he would get back to him in a couple of days.

An hour later, he suddenly remembered that he had run into the next door neighbour of his rental property a week earlier. They were a couple in their fifties. The wife had very good taste in interior design, and had decorated their townhouse quite nicely. They bragged about how pretty their décor looked, and invited him to go inside their house to take a look. It was hot outside, and they had their central air-conditioning on. As Stephen walked inside their house, he felt that the ground floor was cold to the point of freezing. It got somewhat warmer on the second floor, and the temperature was just about right on the third floor where the bedrooms were. He never thought about this weird differential in temperature later. However, now that his tenant had complained, it just dawned upon him why that was. It could simply be explained by a general scientific principle that "hot air rises, and cold air sinks". To keep their bedrooms on the third floor at a comfortable cool temperature, the couple had to turn their air-conditioner on full blast. The cold air sinks down to the lower floors, making the ground floor very cold.

Now that he had figured out the cold air phenomenon, he picked up the phone, and called his tenant. He told him to close all the air vents (where air comes into the rooms) in the first and second floors, and open only the air vents in the third floor. The cold air on the third floor would eventually sink down to the second and first floor, rendering the temperature of the whole house quite uniform.

The idea worked, and the tenants were happy. Stephen was glad that simply knowing a basic scientific principle helped him solve a problem that he had no experience in.

Thus, understanding some fundamental scientific facts can be beneficial in some unfamiliar situations. To the contrary, sometimes not knowing some basic scientific facts can lead to disasters, as the next example will show.

Example 5 Hardwood flooring

A family in Hong Kong wanted to move to a larger place. Eventually, they bought the ground floor of a three-story house. Each floor of the house formed a separate unit with their own individual entrances. As the previous owner of the ground floor unit had vacated the premise for more than a year, it was somewhat in a rundown condition. So the new owners hired an interior designer, Shirley, to supervise a general contractor, Master Chu, to tear the walls down, and have the whole floor re-built. Master Chu had never gone to any trade school, but he learnt his profession hands-on, and was a very skillful person.

The floor measured about 3000 square feet, and the new owners wanted to install quality grade hardwood flooring. The wood planks, each measuring 4.7" × 34.6" (12 cm × 88 cm), had to be shipped from Europe, and cost about HK $100,000 (approx. US $12,800) in total. They arrived in late November, and it was already cold in Hong Kong. The electric wiring and the heating in the premise had still not been installed yet. But since Master Chu was much behind schedule, he went ahead and installed the flooring first. After a few days' work, the floor was beautifully done. Another four months later, the whole residence was completely finished, and it looked superb. Shirley had done an excellent job designing it.

Then summer came, and the temperature rose to 20° C. To the horror of the homeowners, every piece of the hardwood plank buckled transversely. The flooring had been installed in the winter. In the summer, each piece of hardwood board heated up, and expanded. It had no where to go but warped along its width. (The hardwood planks did cool back to their normal shape when the air-conditioning was turned on.)

It seemed that neither Shirley, nor Master Chu knew anything about the basic scientific fact of thermal expansion and contraction. Material increases in volume when heated, and decreases in volume when cooled. Not recognizing this crucial premise had brought upon a mishap.

Example 6 Credit card damage

Liz carried her credit cards in her wallet which she put inside her handbag. A credit card has a magnetic stripe capable of storing data that can be easily read by a card reader.

At one time, she found that card readers had problems reading several of her cards. Merchants had to swipe her cards several times before they could be read, or eventually had to type the credit card numbers in manually. So, she called the credit card companies and had all her cards replaced by new ones. A couple of months later, she experienced the same problem — card readers had difficulties reading her cards.

She wondered what had happened. She knew that the magnetic stripes of credit cards could be damaged if they were put too close to magnets. But she did not think that she had placed them near any magnets. Then she remembered through her high school Physics that an electric current could generate a magnetic field. But what electric equipment could have damaged her credit cards?

As it happened, she joined a fitness club a few months ago. There was a notice on the door of the change room, telling the clients not to put valuables in their lockers even though they put locks on them, as the management had been notified that some locks had been tampered with and some money had been stolen. So Liz put her wallet with her credit cards inside a small bag, which was left right beside the fitness machines when she did her exercises. She thought most likely it was the electric currents in those machines that generated a magnetic field that damaged her credit cards.

After realizing that might be the reason, she again ordered new cards. She would then leave her small bag with the credit cards inside about half a metre away from the fitness machines. From then on, she did not experience any problem of having her credit cards read.

Example 7 Life philosophy

Parenting is difficult. Most parents would fumble along, hoping their kids would turn out alright — behaving morally in the society and interacting with other people in an appropriate manner.

The Smith family has two kids — a daughter, and a son. They left home for university at the same time. About a month later, both ran into some kind of personal problem, and they called their parents for advice. The parents tried to help them resolve their problems over the phone.

Later in the evening, the father sat down at the computer, and wrote the children an email, suggesting what were the general guidelines for common behaviors. The email, with the heading "Life Philosophy", ran as follows:

> *"Unlike actions in the physical world, there are no laws that we can follow to guide our behaviors in life. There is no Newton's Law of motions as in Physics, nor Mendel's Law of Heredity as in Biology. In that case, how should we behave in the forever-arduous path in life that has so many hurdles?*

There are two guidelines. Firstly, one should be considerate and 'Do unto others as you would like others do unto you'. For example, you should love your parents, brothers and sisters, your spouse and your children, as you would like to be loved in return. Secondly, do things in moderation. For example, watching one hour TV (television) per day would be entertaining. However, watching eight hours TV per day would be excessive. Unfortunately, moderation is not well defined. It varies from individual to individual, and depends on our own judgment. The criterion is to balance your actions such that certain action of yours does not drastically affect your physical and emotional well-being, your career, and your relationship with other people. So, if someone spends 16 hours a day working in his office, and ignores doing exercise and any social life, he is not considered to be leading a healthy lifestyle. Similarly, if a person spends half of his salary buying clothing, he would not have enough to pay rent, food, etc.

Thus, one should act and behave in a manner that should be considerate to other people, but not over-indulging oneself in a particular action.

We hope this would help you solve some of your personal problems.

Love always,

Dad and Mom"

Understanding some general principles, especially some fundamental scientific theories, can guide us in unfamiliar territories, proffer solutions to problems that we have no experience upon, and avoid adversities that can possibly happen.

Chapter 9

Alternative Solutions

While there are various ways to view a problem situation, and thus define a problem differently, there are also different ways to solve a problem once it is defined. Some of the solutions may be better than others. If we have the option of not requiring to making a snap judgment, we should wait till we have come up with several plausible solutions, and then decide which one would be the best. How do we know which solution is the best? We will discuss that in the chapter on Probable Value. Generally speaking, we should train ourselves to come up with a few suggestions, and weigh the pros and cons of each resolution.

Let us do a mental exercise. Take a plastic lotion bottle with a pump that has a saddle head dispenser. When the lotion bottle is almost empty, no lotion can flow into the dip tube inside even when we press on the pump dispenser. What would we do? The obvious solution is simply throwing the bottle away, as most people would do. After all, there is probably about 5% of the lotion left at the bottom of the bottle. If the lotion is somewhat viscous, then there may be some stuck onto the inside side of the bottle. So, if we throw the bottle away, there will be at most 10% of the lotion that would be wasted. However, we may want to make the most of our resources, and use up the very last bit. After all, if we can easily save 10% of the resources in the world, why wouldn't we do it?

Basically, we would like to try to get as much lotion as possible out of a lotion bottle with a pump. One possible way is to turn the bottle upside down, and let it lean against a wall. The lotion will slowly drip from the bottom to the top (which is now the bottom). So, every time we want to use some lotion, we can unscrew the pump head, and pour some lotion out. This procedure may be somewhat clumsy. So, are there any alternative ways that we can get the most lotion out of a lotion bottle that has a pump? Let's think about it for a while before we look at some of the answers at the end of this chapter.

In the meantime, let us take a look at another problem, where an alternative solution would have cost much less effort, and better remuneration.

Example 1 Selling a house

In May 1982, Pierre and his wife immigrated to Ottawa, Canada from France. They rented an apartment, just as they had been renting before. Half a year later, Pierre got an inheritance from his uncle. So he talked to a real estate agent about wanting to buy a townhouse to live in. The townhouses in Ottawa at that time were priced at about $50,000. The price was quite reasonable as the mortgage rate had gone up to a whooping 20%. The house market was thus quite depressing.

On one Saturday, the real estate agent showed the young couple two townhouses. Pierre and his wife liked both of them, and made an offer on one. That evening, Pierre thought they still had some spare cash after buying one townhouse, and could easily put a down payment on the other one as an investment and rent it out. So he called the real estate agent, and put an offer on the other townhouse as well. Both offers were accepted, and the couple suddenly realized that they had bought two houses overnight. And that was one big step for someone who never owned any house before.

They assumed the mortgage of the seller for the investment townhouse. Fortunately, the assumed mortgage rate was only 9%, and the interest could be covered by the rent. They were also lucky that the mortgage rate in Canada started to drop, and the house price in Ottawa began to pick up. Four years later, the price of the townhouse had gone up to approximately $65,000. At that point, they wanted to sell the investment townhouse to make a quick profit.

They notified the tenants that they wanted to sell, and the price would be $65,000. The tenants asked whether the couple could wait for about ten days before listing the townhouse in the market, as the tenants wanted to check with the bank for financing to buy that townhouse. The couple agreed. The bank told the tenants that they had to put down at least 10% of the house price as down payment, and that worked out to be $6,500. Unfortunately, the tenants could only come up with $3,500, and they were $3,000 short. Reluctantly, they told their landlords to have the house listed.

More than ten groups of potential buyers went to see the house, and every time the tenants were asked to get the house cleaned, and tidied. Eventually, the tenants were so fed up with the cleaning that they left the house somewhat messy. As a consequence, the landlords did not sell the house at as good a price as they would expect. After more than a month on the market, they finally sold the house at $63,500. They had to pay the real estate agent 5% commission, which worked out to be $3,175. The net selling price of the house after commission is thus $63,500 – $3,175 = $60,325. As the house price when they bought four years ago was $50,000, they made a total profit of $10,325. As their original down payment for the house was $13,000, they wound up making a reasonably good profit of approximately 16% per year for four years. (Compound interest rate formula was used.) And that was a pretty good investment.

Half a year later, Pierre's brother came to visit from France. Pierre bragged to his brother that, without any experience in the real estate business, they made a 16% profit per year by investing in a townhouse. His brother listened, and then said, "Why didn't you give your tenants the money?" Pierre said, "What?" His brother then repeated that why did Pierre not give their tenants the $3,000 that they were short of. He further explained. Pierre should have given the tenants $3,000. That way he could have sold the house to the tenants for $65,000. If this were the case, Pierre would not have to pay the commission fee to the real estate agent and the net selling price of the house would be $65,000 – $3,000 = $62,000. Since the house price was $50,000 when it was bought four years ago, they would have made a total profit of $12,000, which would be $1,675 more than the profit of $10,325 that they now had made. This way, they would not have to go through the effort and trouble of listing with the real estate agent, as they already had a potential buyer. Furthermore, they would also make their tenants happy, as the tenants now would own their own house.

As we will see later on in the chapter on Probable Value, we should always choose a path that has the least effort and provides the most reward, as well as a high probability of success.

Pierre had never thought of the idea that his brother had suggested. Now that his brother mentioned it, Pierre did not think he himself was that smart after all. He and his wife had learned their lesson. They realized that there might be alternative ways to get things done that might have improved the situation. From then on, they would spend time thinking about all possibilities before making any decisions. As years went by, they got smarter and smarter by the year. Twenty-two years later, they would like to move to a single house in an affluent neighbourhood, where the average house price was half a million dollars.

Example 2 Buying a house

Pierre and his wife had only $2,000 in their bank account, not enough to buy a car, let alone a house. But, then, of course, there was always the bank where one could seek financing. They talked to the bank, and the bank would loan them a maximum of $350,000. That, of course, would not be enough to buy a half a million-dollar house. But they came up with several plausible solutions:

(1) Sell townhouse first. The townhouse that they were living in was now worth $220,000. They could sell their townhouse first before making an offer on another house. But the problem was, there was always a possibility that they might not be able to find a house in the affluent neighbourhood that they liked, and they liked their own townhouse. There was no point in moving to a house that they did not like from a house that they liked. They could, of course, sell their townhouse first, and then rent another townhouse, while they waited until they found a house that they liked. However, moving was stressful, and they would not want to move more than it was necessary.

(2) Conditional offer. They could make an offer to buy a house, on condition that they could sell their townhouse. However, the Ottawa house market, especially in that affluent neighbourhood, was hot. Houses in that area, once listed, was sold in one or two weeks. Occasionally, the buyers would get together and bid on a house where offers had been made from several

buyers. Thus, under that market condition, no seller would accept any conditional offer.

(3) Bridge financing. A bridging loan from a bank can be used to provide funds to bridge the gap between the purchase of a house and the sale of an existing house. However, the bank told them bridge financing could be arranged only for a two-month period, provided that they could present to the bank both the offer to buy a house as well as the offer of someone buying their house. As they might not have an offer to buy their house, bridge financing would not be feasible.

(4) Home Equity line of credit. The home Equity line of credit allows the homeowner to get access to a loan up to 75% of the appraised value of his home.

Of all the above suggestions, Pierre thought that only the last one would provide a practicable solution. They would buy a house, using a home equity line of credit on their existing townhouse as down payment. They would then get a home equity line of credit of the new house to pay for the remaining payment of the new house. That is, they would borrow from the new house (which they had not paid for and therefore technically not theirs yet) to pay for the new house. The bank agreed that Pierre's plan was doable. As a matter of fact, Pierre also had a contingency plan. In a worst case scenario that their townhouse was not sold, they could rent out their new house, and he figured that the rent could cover the mortgage interest.

Pierre and his wife then went looking for a house. After a few weeks, they bought one within 24 hours after it was listed in the market. The price of the house was $450,000. They paid 25% of the house price with the home equity line of credit of their townhouse, and the remaining 75% with the home equity line of credit of the new house. That is, they bought the new house without putting a penny down.

They then listed their townhouse in the market and had it sold in half a month after the closing of the new house. Had they use the bridge financing method, they would have been stuck, and would not be able to pay for the new house. In this particular case, the idea of the home equity loan method made it possible for them to purchase and eventually move to a new house.

When time permits, do not rush to pursue a solution that you first come up with. Spend some time thinking about other avenues. Sleep on the problem. Most ideas take time to grow and develop. This incubation period is important for certain concepts to take form. Inspiration sometimes occurs when one is not even thinking about the problem.

Example 3 Brushing teeth

Karen goes to her dentist for a regular check up and cleaning of her teeth twice a year. When she was young, she did not know how to take carc of her teeth. Now, in her early fifties, her gums were receding due to her improper care.

Her dentist then taught her how she should be brushing her teeth. She should not be brushing her teeth back and forth as that would cause the enamel to be scrubbed away, as well as causing the gum to recede. Instead, she should place the toothbrush at a 45-degree angle at the gumline (where the gum meets the teeth), and move the toothbrush from the gum toward the edge of the teeth, so that the dental plaque will be moved away from the gumline. After brushing all the outer teeth surfaces, she should do the same for the inner teeth surfaces.

Karen dutifully followed her dentist's recommended technique. However, she found that, while brushing the outer teeth surfaces was not a problem, it was difficult brushing the inner teeth surfaces, especially the ones on the right-hand-side.

With some practice, she was able to make some improvement in brushing the inner teeth surfaces for those teeth that were located in the centre and on the left-hand-side. But she still found it hard to brush the inner teeth surfaces for both the upper and lower right-hand-side teeth. She mentioned her problem to her dentist, but he did not make any comment.

One morning, before she got out of bed, it suddenly occurred to her what could be a solution. She was right-handed. That was why she found it hard to brush the inner surface of her right teeth. So, she trained herself to use her left hand to brush those particular areas. The idea worked, and she was able to brush all her teeth the proper way with no more problems.

Example 4 Local currency

The Prentices live in San Francisco, USA. In July 2007, the family of four went to tour Prague in the Czech Republic and stayed in a hotel for four nights. The travel agent had booked the hotel for them, at the rate of 83 Euro per room per night.

When they arrived at the hotel, the mother suddenly thought whether they could pay the hotel in crowns, which is the local currency of the Czech Republic. She asked the front desk clerk, and was told that indeed she could pay in crowns, and the hotel rate would be 2347 crowns per room per night. She then told the clerk that they would pay them in crowns.

As one Euro was about 30 crowns, she just saved herself $(83 \times 30 - 2347) = 143$ crowns per room per night by paying in crowns. As they rented two rooms for four nights, she saved herself (143 × 2 × 4) = 1144 crowns, which was approximately 56 US dollars.

Example 5 Plastic laundry hampers

The Lees live in Seneca Falls, a small town in New York State in USA. In the year 2002, their son, Peter, left home to attend university at New York City. Every three months or so, the parents, Emanuel and Lisa, would drive for about five hours to go visit their son, and see how he was doing.

In November 2004, the parents drove down to New York City to see Peter. During lunch, Peter told them that his plastic laundry hamper was broken a month ago, and he went to buy a new one. (A hamper is a large basket and usually has a cover. Plastic laundry hampers usually have rows of holes on the sides to vent moist or smelly clothing. Each hole measures approximately 0.4" × 0.4" (1 cm × 1 cm)). Since he did not have a car, he had to walk twenty minutes to carry it all the way home. The hamper was not heavy, as it weighted only about 3 lb (approx. 1.7 Kg). However, it was large since it measured 19" × 14" × 24" (48 cm × 36 cm × 61 cm), and it was somewhat awkward to carry. He had to hold the hamper in front of him while walking home.

The father then asked him why did he not ask the store clerk for a plastic bag. (Or better still, to be more environmentally friendly, Peter should have brought along a piece of cloth.) As the hamper has holes on its sides, Peter could have threaded the plastic bag into the holes, tied a knot, and made a handle out of the bag. He could then carry the hamper by his side, making it much easier for him to carry it home.

The father then told his son that when he went to an out-of-town university thirty years ago, he had to buy two metal bookshelves at one point. Each bookshelf came unassembled in a paper box that measured about 36" (91 cm) in length, 8" (20 cm) in width and 3" (7.6 cm) in height (thickness). He had to walk twenty-five minutes back home, and carrying those two boxes would be rather clumsy. So he brought along with him a small pair of scissors, and he asked the store clerk for two paper bags that came with rope handles. (Back then, paper bags were used instead of plastic bags.) A paper bag, when open, looks like a box without a lid. The paper bag measured about 17" (43 cm) in length, 6" (15.2 cm) in width and 15" (38 cm) in height. Using the pair of scissors, he cut open the four edges that joined the four vertical sides, and then folded the two narrower sides onto the bottom of the bag. He repeated the same procedure with the other bag, and then laid one on top of the other on the floor. The two bookshelf boxes were next laid side by side on top of the bottom of the cut-open bags, with their lengths parallel to the length of the bags, and their heights parallel to the width of the bags. He then grabbed the four handles of the paper bags. After checking that the boxes were well balanced from the bottom of the bag, he took the two metal bookshelves home without much difficulty.

The father then continued talking to his son about how his own father took two large bottles of cooking oil home after walking half an hour. His family was poor when he was a kid, and his parents had to watch for bargains when it came to shopping groceries. One weekend, his father walked by a store, and cooking oil was on sale. So he bought two large bottles. Each bottle contained 3 litres of oil and weighed about 3 Kg. The store clerk tied the two bottles together with a piece of string to make them easier to carry. After carrying the two bottles with the string for about five minutes, his father felt that the string was cutting into his hand. He figured that he needed some cushioning for his hand to reduce the pressure from the string. So he took off both his socks, put one sock over his right hand, and put the other sock over the first sock. He then carried the two bottles with the string using his double-socked right hand, and walked the rest of the way home.

The moral of the lesson, Emanuel told his son Peter, was that always try to look for alternative ways to get your task done easier, and make your life more pleasant.

Example 6 Cooking oil bottle

In the Smith family, Albert does the grocery shopping, and his wife Hilary does the cooking.

One day, Albert bought two bottles of cooking oil of a certain brand, as it was on sale. When Hilary got around to using one bottle, she soon found out that there was a problem. The opening of the bottle had a large diameter of 1¼" (3.2 cm), and it was very difficult to control the amount of oil pouring out of the bottle. When she poured the oil into a pot or a wok, she always wound up pouring much more than she intended to. So she told Albert never to buy this brand of cooking oil again.

Albert agreed that the opening of the bottle was too large for the purpose. One would think that there might be a conspiracy of the manufacturer, as the consumers would wind up pouring more oil than necessary, and wind up having to buy more oil after. But Albert had an idea.

He took a small piece of aluminum foil and wrapped it over the bottle opening. He then used a twist tie to tie the piece of aluminum foil around the neck of the bottle so that the foil would not fall off. After securing the foil, he poked a hole of 0.1" (2.54 mm) diameter in the foil. The small hole restricted the flowing out of the oil, and rendered the pouring much more controllable.

Hilary was quite happy with the solution, and did not complain any more.

Example 7 Toilet training

When a child is a baby, she usually wears diapers. Toilet training is teaching her to grow out of wearing diapers, and use the toilet.

Charles and Betty both work in the daytime. They have a daughter, Jackie. When their daughter was just born, Betty stayed home with her for half a year. As she had to go back to work after, she had to give her to a babysitter to take care of her during the day.

When Jackie turned two, Betty thought it was time to toilet train her. As Jackie was a smart girl, she learned fast. And she did not have any problem going to the toilet by herself. But, unfortunately, she would wet her pants every night after she had gone to bed, in spite of the fact that Betty had taken her to the toilet before putting her to sleep.

Charles and Betty did not have any idea what they could do to stop Jackie from wetting her pants at night. So they consulted the babysitter. The babysitter, apparently having a lot of experience with kids, had a simple suggestion. As the parents usually go to bed a couple of hours after the kids, she asked them to wake Jackie up to use the toilet before they went to bed. Small kids usually can fall back to sleep right after.

Her advice worked, and Jackie did not wet her pants at night anymore.

This example just shows that it is a good idea to talk to other people, especially experts in areas that one may not be familiar with. However, that does not mean that one should slack off doing one's thinking, as the next example will show.

Example 8 Mortgage penalty

In the year 1991, George moved to Cornwall, Canada. A year later, he bought a house, and had to borrow $100,000 mortgage from the bank. He was told that if he had to terminate his mortgage early, he had to pay three months interest penalty of the remaining principal of the mortgage. He was also told that he could pay off 15% of the original principal amount of the mortgage, penalty-free, each year.

In the following years, he would pay off a couple of thousand dollars in some years when he got spare cash. In early 2006, he had to sell his house and move to Toronto. After the house was sold, he had to pay back to the bank what he still owed. At that point, he owed the bank $40,000 mortgage, and the mortgage interest was 7% a year. The three months interest penalty would be $700. However, he knew there was an alternative way to save himself some money. He could pay off 15% of the original principal amount of the mortgage, and then pay the three months interest penalty of the remaining mortgage. And that was what he instructed his financial advisor at the bank to do. So, he wound up paying only $437.50 interest penalty and saving himself $262.50. To him, it was obvious that was the way that it should be done.

A few months later, he was somewhat surprised when he read it in the newspaper that many mortgage holders did not know that they had a choice, and a lawyer had initiated a class-action lawsuit over mortgage penalties with several Canadian banks. The suit claimed that the banks overcharged the mortgage holders by not telling them that they had an option of paying fewer penalties when they repaid their mortgage in full prior to maturity. As a matter of fact, the lawyer succeeded in settling the case with one of the banks. The particular bank would pay back part of the penalty difference to its customers. However, the other banks fought back by saying that it was not their obligation to tell their customers that they could save some money by terminating their mortgage differently.

Whether the banks have such an obligation is somewhat debatable. However, the bottom line is, the financial advisors at the banks work for the banks, and they are supposed to bring money in to make the banks profitable. As a customer, your objective is quite different from that of the financial advisors, who have conflicting interest with you. Therefore, you should try to get as much information as possible from the advisors, and think for yourself if you actually want to save yourself some money.

Example 9 Microfiche

Microfiche is a flat sheet of microfilm containing a miniature photographic copy of printed or graphic material, e.g., a document, newspaper, etc. The image of the original material is usually reduced about 25 times for easy storage. It can be provided as a positive or a negative, and more often the latter.

Tracey lives in Macau. One day in the year 2006, when she was looking through some old documents in a paper box, she found an envelope with the words "Land deed" and the name of her deceased grandfather written on it. She figured it was the deed of a piece of land that her grandfather had bought in China more than eighty years ago.

She opened the envelope and found a piece of microfiche negative that measured 4" × 6" (10 cm × 15 cm). She very much wanted to know what was written in the microfiche. So, she took it to several photo development stores to see whether they could develop the negative. However, they did not have the facilities to do so. Eventually, she found one store that could do it, but it would cost about US $100. She thought that was somewhat expensive and she told the store manager that she would think about it, and get back to him later.

Back home, she met her brother, and talked to him about developing the microfiche. Her brother suggested that she could give the microfiche to their nephew who was familiar with computer programming. He could scan the microfiche using a scanner, and then enlarge the digital image using computer software, and then print the enlarged image out in several sections.

Tracey followed her brother's advice, and gave the microfiche to their nephew, who later printed the document out in several pages with no problem.

Example 10 Horizontal venetian blind and toilet paper roll

A horizontal venetian blind is a window blind made of long narrow horizontal slats held together by strings. The slats are opened or closed to admit or exclude sunlight by rotating a drive rod.

Jacques and Simone had a horizontal venetian blind installed in their bedroom. Simone always opened the blind in the morning to let the sunlight come in, and closed it at night before they went to bed. Jacques noticed that when Simone closed the venetian blind, the slats were always tilted down. One day, he asked her whether there was any particular reason why the slats were tilted down when the blind was closed. Simone answered that there was no specific reason. She never thought there was any difference whether the slats were tilted up or down.

There was, Jacques told her. When the slats were tilted down, the glare from the morning sun could wake them up earlier than they might intend to. Furthermore, the afternoon sun will heat up the room faster in the summer and make the room hot. In addition, if the slats had been in the down position for a long time, the carpet right beneath the window could get faded by the sunlight. When the slats were tilted up, almost all the sunlight would have been blocked.

The manner that a task gets done may have made a difference in the consequence. However, some people are not aware that there may be a distinction. Incidentally, there is this age-old philosophical question whether toilet paper roll should roll over the top, or should roll under the bottom when the roll is placed in a wall mount toilet paper holder.

Some people would not care less, and put in a new roll of toilet paper when the old one runs out, without giving it too much thought. However, there is a certain advantage in placing the toilet roll in the "over" position than the "under" position, as the former manner makes the toilet paper more easily accessible, especially in the dark when one has to go to the bathroom in the middle of the night. Nevertheless, if there are pets or kids that like to unroll toilet paper, the "under" position would be preferable, as it makes the toilet paper more difficult to unroll.

Simone did not realize this important difference. When she first got married to Jacques, she would randomly put the roll in an "over" or "under" position. As they did not have any kids or pets, Jacques thought that the roll should be in the "over" position. So every time he saw an "under" roll in their house, he changed it back to "over". Eventually, he had a word with Simone, explaining to her the convenience of having an "over" roll, and won her over to the "over" camp.

Example 11 Apartment renovation

Janice lives in Shanghai, a few blocks from a fifteen-year-old apartment that her mother had just bought. The apartment measures 900 square feet (84 square metres). It is on the twelveth floor facing the harbor and has a beautiful view. However, her mother did not like the layout of the apartment, and would like to tear down the walls and do a complete renovation.

As Janice is an interior designer, she offered to design the layout, and supervise the contractors to do the renovation. After the renovation was completed three months later, her mother moved in.

Four months later, Janice's brother, Alexander, came visit his mother from out of town. When he got into her apartment, he was surprised to see that a built-in wardrobe closet was blocking part of the window facing the harbor. When he asked Janice why was the wardrobe closet put in that particular location, he was told that their mother wanted a total of 12 feet (3.7 metres) long wardrobe space to put all her clothing. So she designed a set of two parallel 6 feet (1.8 metres) wardrobe closet with a corridor in between to access the clothing. She believed that was the best location to put the closets, and it just so happened that one of the closets was blocking part of the window.

Alexander found her design concept rather bizarre. One of the assets of the apartment is that it has a view facing the harbor. Blocking part of the view is somewhat odd. In addition, the corridor in between the closets is wasting valuable space.

Janice does not seem to understand what the constraints of a problem are. In this particular case, the constraint is a 12 feet long wardrobe. The wardrobe can be put anywhere in the apartment as not to consume valuable space, and not blocking any view.

Every problem has constraints. The usual constraints are time, money and labor. We do not have infinite amount of time, nor unlimited financing, nor extensive labor to perform a task. Other constraints are rules and regulations that we need to follow. Within the guidelines, we should figure out how we can maximize our gain. The mortgage penalty example described earlier presents a good illustration of how this can be done. We just have to work within the constraints to get the optimal solution. Some problems can have quite a few answers, in spite of the various restrictions. It is up to us to find the most desirable and satisfactory resolution.

In summary, we should always look for alternate solutions, and train ourselves to think outside the box. The box is the normal paradigm or pattern that we are accustomed to.

Thomas Kuhn, in his book *The Structure of Scientific Revolutions*, describes a scientific paradigm as a scientific theory or a pattern of looking at the world that has attracted a group of followers. A paradigm shift would occur when there is a change in basic assumptions within the ruling theory of science, thus leading to a scientific revolution.

Similarly, we can look for different ways to solve a problem, and avoid the traditional paradigm that is being viewed, and arrive at completely different answers. As the nineteenth-century German philosopher Nietzsche once wrote: "To build a new sanctuary, a sanctuary must be destroyed. That is the law".

Furthermore, we may need to realize that in certain problems, some of the constraints can be lifted. We may also have built-in assumptions that may have obstructed our finding a different solution. Those assumptions have to be eliminated to allow new ideas to germinate. We do not necessarily have to follow a certain mode of thinking. In problem solving, the guideline can be: there is only one absolute, that there is no absolute.

Now, let's go back and take a look at various ways to get the last bit of lotion out of a lotion bottle.

9.1 Lotion bottle with a pump dispenser

Here are some of the alternative ways to get as much lotion as possible out of a lotion bottle with a pump: -

(1) Lay the bottle on its side on the table, and let the lotion settle in the inside of one side. Every time you use it, unscrew the pump head, and scrape some lotion out with the dip tube. However, this method looks clumsier than the original suggested method of turning the bottle upside down.

(2) Find a small empty round facial cream bottle. Take its cap off. Remove the pump head of the lotion bottle. Turn the lotion bottle upside down and let it sit on the open cream bottle, allowing the lotion to drip into the cream bottle. After the dripping is finished, you can then use the lotion off the cream bottle. There will be still some lotion sitting in the inside of the lotion bottle near the opening at the top. You can use your finger to scrape it out if you want to use this last bit.

(3) Some lotion bottles come in two different packaging — one with a pump dispenser, and the other with a dispensing flip-up cap. The latter is usually slightly cheaper than the former. Purchase both kinds. When no more lotion can be pumped out from the bottle with a pump dispenser, exchange the pump dispenser with the dispensing flip-up cap. Then turn the almost empty lotion bottle upside down and have it lean against a wall, letting the lotion drip down to the opening. Every time you want to use some lotion, squeeze the bottle such that the lotion

comes out of the flip-up cap, while holding the bottle upside down.

(4) Cut the lotion bottle into two halves with a knife. You can then use all the lotion to the very last bit.

Chapter 10

Relation

Relation is the connection and association among different objects, events, and ideas. Problem solving, quite often, is connected with the ability to see the various relations among diversified concepts.

As an exercise, we can pick an object and think of what other uses that we can dream of. Imagine as many applications as we possibly can, no matter how far-fetched they can be. Try this for a piece of brick, a paper clip, a pencil, etc.

Let's take an example of a drawer. It is usually considered as a boxlike container in a piece of furniture, and can be made to slide in and out. But, can we think of other situations that we can use it for?

In a British university student residence, some students were trying to prepare a sit down dinner party of ten people, to celebrate one of the students having awarded a scholarship. But there were not enough chairs. So someone came up with the idea of using the drawers of the desk. Each drawer was set vertically with the bottom of the drawer on the floor; the length of the drawer was just about equal to the height of a chair. The students were then able to sit on the standup drawers, and enjoyed their dinner.

A problem may not be solved unless the person is able to see the combination of two or more than two seemingly unrelated concepts, as the following several examples will show.

Example 1 Masking tape

The masking tape is used mainly in painting — for masking off areas that should not be painted. What other uses can it be used for?

Joseph did not take good care of his teeth when he was a child. Now that he was in his late forties, his teeth started deteriorating. His dentist told him that he should floss and brush his teeth every time he finished eating anything. Since he had three meals and three snacks each day, he had to dental floss six times a day. As he had to wind the floss around the first joints of each index finger, the joints got somewhat chafed after a few weeks of flossing. In order to protect the skin of his fingers, every time before he started flossing, he would tear a small piece of masking tape, and wrap it around the first joint of each index finger. From then on, the skin of his index fingers did not get chafed.

Masking tape is convenient to use, as it does not need to be cut with a pair of scissors. It can simply be torn off with one's fingers. It is quite a versatile material that can be adapted for different purposes, as the following example will demonstrate.

Example 2 Drill bit

A drill is a tool with a rotating drill bit for drilling holes in wood, metal, etc. At the end of the drill is a three-jaw chuck for holding the drill bit. The rotary action of the chuck moves the jaws in or out along a tapered surface. The taper allows the jaws to incorporate various sizes of drill shanks. The chuck grips the drill bit tight when the chuck body is turned manually or by using an auxiliary key.

After the drill bit is tightened securely in the chuck, it is pressed against the target material and rotated. The tip of the drill bit cuts into the material by slicing off thin shavings. To make a hole, one should start with a drill bit of small diameter, and then slowly increase the diameter of the drill bit until the desired diameter of the hole is reached.

Richard was doing some renovation in his house, and he asked his friend John whether he could come over and help. John gladfully agreed.

Richard had to drill a hole in a piece of metal. He started off with a titantium drill bit of 1/16" (1.6 mm) diameter. He then slowly increased the diameter of the drill bit to ¼" (6.4 mm), which was the size of the hole that he wanted. At that point, he found that the drill bit was slipping against the jaws of the chuck. It was probably caused by the hole putting up a large torque resistance. Richard would like to get the task finished, but he did not exactly know what to do. So, he asked John whether he had any suggestion.

John asked Richard to take the ¼" drill bit off the chuck. He then tore off a small piece of masking tape of 0.2" × 1" (0.5 cm × 2.54 cm), and told Richard to tape it longitudinally onto the drill shank, and put the drill bit back into the chuck.

Richard did what John suggested. The friction furnished by the small piece of masking tape provided the necessary friction for the jaw of the chuck to grip the drill shank. And Richard quickly finished his task with no further problem.

Example 3 Soft drink and pants

The parents took their five-year-old daughter and three-year-old son to a pizza restaurant. They ordered pizza and soft drinks. As usual, the soft drinks came first. While they were waiting for the pizza to come, the son tipped the whole glass of soft drink on the table, spilling the drink onto his lap and completely wetting both his polyester shorts and cotton underpants.

What should the parents do? Should they ask the waiter to pack the pizza for takeout and go home? Or, should they go to a nearby shopping mall to get a pair of shorts for their son, and then come back for the pizza? But the pizza would be cold by then.

The father thought for a second. He got up and went into the washroom, and then quickly came back out. He had gone inside to check whether there was an electric warm air hand dryer. He then took his son into the washroom, and took off both his shorts and underpants. After drying the shorts with the hand dryer, the father put the shorts back onto his son, and wrapped the wet underpants with some paper towels to take home later.

All these took about two minutes. By the time father and son came back to the table, the pizza still had not arrived yet. They sat down and enjoyed their meal after, and then went back home.

Example 4 Lost screw

The Disneyland in Orlando, Florida is about 24 hours drive from Ottawa, Canada. Some parents would like to take their kids to Disneyland during holiday seasons. Father and mother can choose to drive 24 hours non-stop from Ottawa to Orlando, sharing the driving between them, and arriving in Orlando dead-tired.

One Christmas, Rod wanted to take his wife, their fourteen-year-old daughter, and their twelve-year old son to Disneyland. As both he and his wife were not excellent drivers, they chose to take turns and drive eight hours per day, thus taking three days to drive to Orlando. They spent about two weeks in Disneyland and had a very good time.

On the day that they were heading back to Ottawa, they woke up late, and by the time that they got everything packed, it was already eleven in the morning. Rod started his car and put on his prescription sunglasses. He then found out that the screw that held the left leg to the eyeglass frame had fell off, and was nowhere to be found. Driving without sunglasses for several hours under the sun would be quite uncomfortable, especially when there was snow on the ground reflecting the sunlight. He could, of course, looked for an optical store, and got it fixed. However, he would like to leave Orlando as early as possible in order to avoid driving in the dark in the evening.

Rod was wondering what could he use to temporarily get his sunglasses fixed, so that they could get on their way as early as possible. He quickly took a piece of dental floss, and threaded it through the holes where the screw was supposed to go, tying the leg to the frame with a knot. As he had no scissors with him, he used his nail clippers to cut the ends of the dental floss off. The whole process took a few minutes, and then the family drove happily toward Ottawa.

In this incident, Rod observed in his mind what instrument might be useful to get his problem solved quickly. He scanned through his database to find out what is relevant to the problem-situation. He saw the relation between the screw and dental floss. He then tried out his idea, and got the problem solved.

Example 5 Icy driveway

In Canada in the wintertime, it snows quite often. Occasionally, there is also freezing rain. Freezing rain develops when falling snow comes upon a layer of warm air deep enough for the snow to melt and turn into rain. As the rain continues to fall, it passes through a layer of cold air, and is cooled to a temperature below freezing point. However, the rain does not freeze — a phenomenon called supercooling. When these supercooled rain droplets hit the frozen ground, they instantly freeze, forming a thin film of ice, called glaze ice. Glaze ice is very smooth and provides almost no traction, causing vehicles to slide even on gentle slopes.

It was December 23, 2006 in Calgary, Canada. Outside temperature was –20° C. The freezing rain had finally stopped at about 10 pm. About an hour later, the teenager daughter came home from a party. She told her father that she left the car parked on the road, as there was no way she could drive the car back into the garage. The car simply slid back into the road every time she tried.

The driveway is about 30 feet (9.1 metres) long and slopes down at an angle of about 4° to allow water to drain away from the house. Because of the freezing rain that fell in the evening, a layer of glaze ice had formed on the surface of the driveway, which became very slippery. However, the father could not believe that the car could not be driven back into the garage. In any case, he would not want the car to be parked on the road, as it would obstruct the city's snow removing crew clearing the street.

So the father went outside the house and tried to drive the car back into the garage. He soon found out that his daughter was right. It was futile driving the car up the driveway as the car would slide back after getting half way up the driveway.

The father remembered that there were some folded paper boxes stacked on the shelf inside the garage. They each measured 36" × 45" (91 cm × 114 cm). So he grabbed two of them, and laid them along the length of the driveway, one near the end of the driveway and the other near the garage opening. He drove the car up the driveway such that the right tires would roll over the folded paper boxes, which provided the friction that he wanted. He got the car back into the garage with no problem.

Example 6 Digital camera

In the summer of 2006, the Carpenters drove to Berlin, Germany from Amsterdam for a vacation. They would stay in Berlin for five days, and would then head for Prague in the Czech Republic.

While in Berlin, they stayed in a hotel located in the suburb, as the rate would be much cheaper than the hotel rate in downtown Berlin. The hotel is only about five minutes walk to a subway station, from where it will take twenty minutes to go to downtown by the subway train.

The Carpenters preferred to plan ahead. They would like to know the fastest route to get out of the suburb of Berlin and head toward Prague when it would be time to leave the hotel. So they asked the hotel front desk clerks for direction, but the clerks were not quite sure, as they probably did not have cars themselves. The clerks also did not have a map of the suburb of Berlin.

One night, as they came back from Berlin downtown, and got out of the subway exit at the subway station near the hotel, the daughter suddenly noticed that there was a map of the suburb of Berlin on the display window at the subway entrance. It was somewhat dark already, making it difficult to read the map.

Furthermore, everybody was tired, and wanted to walk back to the hotel as quickly as possible.

Fortunately, the daughter got an idea. She took the digital camera, and took a picture of the map. Later on, back in the hotel, she was able to use the zoom feature of the camera to study the map on the 2.5" × 1.75" (6.4 cm × 4.4 cm) liquid crystal display window. She figured what was the quickest way to leave the hotel and head toward Prague. Following her suggestion, they did not waste any time leaving Berlin a couple of days later.

Example 7 Thorny weeds

Brad and Katie had just bought a house. On the weekend, Brad tried to clean up the garden at the backyard by using a 7" (18 cm) long garden shears to cut down some of the bushes. Even though he was wearing a pair of garden gloves, he was cut by the thorns of some tall weeds that were hidden inside the bushes. He then realized those weeds were about 5' (1.5 metre) tall, and their stems were filled with thorns about 0.39" (1 cm) long.

Brad believed that he could cut the weeds into lengths of 2½' (0.76 metre) and then pick the pieces up and put them into a recycled paper lawn bag. But he would need a garden tool to pick up the cut pieces so that he would not be hurt by the thorns. So he went to a couple of hardware stores to look for a garden tool that would allow him to pick up those thorny weeds. However, he could not find any tool that would fit his purpose.

After he came home, he mentioned to Katie about his fruitless trip to the hardware store. Katie simply asked him why did he not use a pair of barbecue tong. A barbecue tong is usually used for picking up food in a barbecue, and has a long handle.

Brad thought it was a good idea. So he switched to using an 18" (46 cm) long hedge shears to cut the weeds and picked the cut pieces with a 14" (35.6 cm) long barbecue tongs. The chore went smoothly and he did not get himself cut again by the thorns.

Example 8 Itchy scalp

Emanuel works in Glasgow, Scotland. He goes back to see his mother in Singapore once a year. In her mid-eighties, his mother lives with a maid, and is in reasonably good health. His sister, Nancy, lives two minutes away from her mother's, and comes visit her often.

In September 2007, Emanuel flew back to Singapore and would stay at his Mom's for about three weeks. During his stay, he heard his mom complaining about her scalp being itchy. Her scalp had started getting itchy about half a year ago. It got so itchy that sometimes it woke her up in the middle of the night. Nancy knew about it and had bought her an expensive bottle of liquid to rub on her scalp, but according to his mom, it did not help.

When Emanuel asked Nancy where she bought the bottle of liquid for their mom's hair, he was told that she bought it from a hair salon and it was a good product. He commented that if the product did not work, then there was no point in continuing with it even though it might be a good product. But Nancy insisted that mom should keep trying.

At that point, Emanuel picked up the bottle and looked at the label. He found out that the liquid was manufactured from plant extracts and was meant for hair loss, and not for itchy scalp. Nancy had never bothered to read the label.

He then saw an inexpensive bottle of body lotion on the table, and asked his mom to try rubbing it on her scalp. He figured that his mom's scalp was just too dry. Body lotion is actually meant for dry skin on the body, but not for the scalp. But as far as Emanuel was concerned, if it did not say that it could not be done, then it could be done. And in any case, he did not believe that there was any harm in trying.

His mom followed his suggestion, and rubbed the body lotion on her scalp. The itchiness stopped. From then on, she used the lotion once in the morning, and once at night before going to bed. And her scalp did not feel itchy any more.

Even half a year later, when Emanuel called his mom from Glasgow, her mom told him the body lotion was working fine on her scalp, and it stopped her itchiness.

Example 9 Air-conditioned restaurants

Paul lives in Edmonton, Canada. He goes to Hong Kong to visit his brothers and sisters once every two years or so. He would try to avoid going in the summer, as the temperature in Hong Kong ranges between 26° C and 34° C and is unbearable for someone who is accustomed to the cold climate.

However, August 2006 was his brother's 50th birthday. So he went back for his birthday party.

The next day after he arrived, Paul met his sister for lunch in a small restaurant across town. It was 30° C and was hot and humid. However, inside the restaurant, the air-conditioning was full-blast to the point that it was freezing. It is quite odd that quite a number of restaurants in Hong Kong have their air-conditioning turned on super-high in the summer, as if they did not have to pay for electricity. Local people quite often bring sweaters with them even though it is extremely hot outside, in case they might have to go to a restaurant.

Paul was not aware of this possible huge contrast in temperature in the Hong Kong environment. He did not bring a sweater with him. And after sitting in the restaurant for five minutes, he was beginning to feel cold. He was wondering what he could do to keep himself warm. Fortunately, he had a document bag with him. The bag measured 15" × 12" (38 cm × 30 cm). So he put his bag against his chest and held it in place with his left arm. In addition, he sat back such that his back was backing on the back of the chair. That way, he was somewhat insulated from the coldness, and that kept him warm during the whole lunch hour.

Paul learns from his experience, and from then on, he always remembers to bring a sweater with him when he goes outside in the hot summer days of Hong Kong.

Seeing a relation among different concepts can quite often lead to creative and unconventional solutions. This is quite often associated with creativity. So, what exactly is creative thinking?

10.1 Creativity

Creative thinking is often thought to have a different mental process from our everyday ordinary thinking. It is said that it quite often arises from inspiration or insight that occurs out of the blue and probably from our subconscious mind. This Aha! Experience would then quickly contribute to the solution of our problem.

Over the past fifty years, psychologists, sociologists and neuroscientists have performed experiments on creativity, and the current consensus is that creative thinking is no different from ordinary thinking. Let us first discuss what ordinary thinking consists of, and then how various aspects of creative thinking can be interpreted in terms of ordinary thinking. Finally, we will describe an example of scientific creative achievement and show that it can be explained by the cognitive components of the ordinary thinking process.

10.1.1 *Ordinary thinking*

The cognitive components of the ordinary thinking process consists of: (1) Memory — remembering of past events or searching of information already stored in the brain (2) Planning — formulating schemes or programs in order to accomplish a certain task (3) Judgment — evaluating the outcomes of several paths of actions (4) Decision — choosing among several plans of action.

Let us take a look at an example of ordinary thinking. Let's say that someone said that she had gone shopping yesterday. In that sense, to her, thinking would mean remembering which shopping centre she went to, and what exactly did she buy.

Of course, ordinary thinking can get more involved. For example, if someone has difficulty trying to open a metal lid off a glass jar of jam, then he would have to choose which path of action he has to take to hopefully get the lid opened. He would attempt to remember his past experience, as well as his experiences of having watched other people open a glass jar in the past. He can (1) wear a rubber glove to give his hand more friction with the lid, (2) tap the metal lid with the handle of a metal knife to loosen the tightness between the lid and the glass jar, (3) turn the jar upside down and immerse the metal lid in a shallow dish of hot water so that the lid can expand. He eventually has to decide which action he will take, and may choose to follow the path that he thinks will involve the least amount of work, and then proceed to other paths if the first action fails. Thus, he follows the ordinary thinking process.

Now, let us assume that this person has never seen anyone open a metal lid off a glass jar by immersing the lid in hot water. However, he suddenly remembers that metal expands more than glass when heated, and thinks that he may use this principle to open the lid from the glass jar. To him, this is creative thinking, as he is going to try something new — something that he has never done before.

These mental processes of creative thinking, as will be shown, are no different from those of ordinary thinking.

10.1.2 *Creative thinking*

Creative thinking introduces something new or different. The creation can be a solution to a household problem, or it can be a scientific discovery or an engineering invention that is earth-shaking. If a person has done something new from his viewpoint, but not new in the perspective of other people, the creativity is local. If a person has done something new in the eye of the whole world, then the creativity is global.

In general, there are several aspects of creative thinking that needs to be clarified.

10.1.2.1 *Knowledge*

Some people think that creativity requires radical ideas that come from nowhere. This is very much far from the truth. Knowledge is extremely important in creative thinking. Some problems, e.g., some of the puzzles, are knowledge-lean, i.e., little or no knowledge is required to solve the problems. This originates from the manners in which they are structured. However, most problems, e.g., in scientific discoveries, are knowledge-rich, i.e., deep knowledge and expertise is needed. Thus, it pays for an individual to find out as much relevant information as possible, as well as discuss with others concerning the problem that she has on hand. New ideas are usually built on combination of existing ideas, or borrowing an idea from an analogous problem situation.

Some people argue that knowledge can be intrusive and can blindside the problem solver into getting into a mental rut, and cannot think outside the box. It will be better off to start with a clean slate. This, of course, can be true in some rare cases, where the person may have been trenched in some built in assumptions that certain previous knowledge has to be right, and would not attempt something new. But, in general, it is almost impossible to solve most problems without studying what other people have done before.

10.1.2.2 *Insight*

A person may be working on a problem and feel like she is not getting anywhere. Then suddenly, an idea comes through and everything is clear. The solution is just right in front of her eyes. This leap of insight is most fulfilling. But is insight any different from the ordinary analytic thought process?

In ordinary analytical thinking, the person would analyze the problem based on her knowledge and expertise, and the problem is solved step by step. She may be able to find an analogous problem that she has solved before. The solution would then be carried out to see whether it is successful. No insight seems to be involved.

The concept of insight, some psychologists believe, plays a certain role in creative thinking, and the process is completely different from that of analytical thinking. They believe that insight is the result of a restructuring of a problem after a period of non-progress (impasse) where the person is believed to be fixated on past experience and get stuck. A new manner to represent the problem is suddenly discovered, leading to a different path to a solution heretofore unforeseen. It has been claimed that no specific knowledge, or experience is required to attain insight in the problem situation. As a matter of fact, one should break away from experience and let the mind wander freely.

Nevertheless, experimental studies have shown that insight is actually the result of ordinary analytical thinking. The restructuring of a problem can be caused by unsuccessful attempts in solving the problem, leading to new information being brought in while the person is thinking. The new information can contribute to a completely different perspective in finding a solution, thus producing the Aha! Experience.

10.1.2.3 *Unconscious mind*

It has been said that unconscious cognitive processing is important in creative thinking. The person is unconsciously thinking about the problem while she is consciously thinking about something else. This kind of unconscious incubation can lead to a sudden illumination, where an unexpected presentation of solution appears out of the blue. Connection among ideas that is beyond the reach of conscious thinking is believed to be made possible by unconscious processing. However, psychological studies so far have shown weak support for unconscious processing in creative thinking. The claim is that the person has actually been consciously thinking about the problem on and off.

Unfortunately, at the present moment, there is no satisfactory model to explain incubation and illumination, which have been reported by a number of scientists in many scientific discoveries.

In summary, current psychological theory has concluded that the mental processes of creative thinking are almost no different from those of ordinary thinking. An example of creative thinking will be used as an example to illustrate this point.

10.1.3 *Double helix*

The discovery of the structure of DNA, the genetic material, can be considered as creativity of a high calibre. Watson and Crick discovered the double-helix model of the DNA structure in 1953.

Biologists had been trying to find the composition and structure of the DNA for more than fifty years. Watson and Crick succeeded in finding the correct model after about one and a half years of work, while other research teams failed to do so after much longer period of dedication. Scientists quite often take different approaches in solving a problem. While some succeed, others do not. The double helix discovery will provide some insight into how certain people utilize their creative thinking to achieve their goal. Furthermore, it will show that the mental procedures of creative thinking are the same as those of ordinary thinking.

10.1.3.1 *Genetic material*

Deoxyribonucleic acid (DNA) is a nucleic acid that contains the genetic blueprint needed to construct other components of cells, such as protein molecules. DNA is found almost exclusively in chromosomes. A chromosome is actually a very long DNA molecule, with associated proteins. However, there is more protein than DNA in chromosomes, and that led to the belief that protein might be the significant material carrying the hereditary information of a living organism. It was only in the 1950's that many scientists began to agree that DNA was the material that carried the genetic information. Both James Watson and Francis Crick realized that DNA was more important than proteins in the storage of genetic data. Presumably, scientists need to know which path to follow in order to get to the right destination.

10.1.3.2 *Watson and Crick at Cavendish Laboratory, Cambridge*

James Watson (1928–) received his PhD in genetics at Indiana University when he was twenty-two years old. At the suggestion of Luria, his PhD supervisor, he went to Europe in 1950 to learn more about the chemistry of nucleic acids as Luria thought that would help Watson understand how genes function. At a conference in Naples in 1951, Watson saw a slide of a DNA molecule produced by X-ray crystallography in a talk given by Maurice Wilkins (1916–2004), who was working at King's College in London. The X-ray photograph fascinated Watson, as that meant DNA was a crystal and had regular structure. The structure of DNA could then be deciphered without possibly too much work. Shortly after, Watson was able to get Luria to help him get an appointment at Cavendish Laboratory at Cambridge University, where he could learn X-ray diffraction.

In September 1951, Watson joined Cambridge University, where he met Francis Crick (1916–2004). Crick learned Physics before World War II, and made use of his knowledge to work in research for the Admiralty Research Laboratory during the war. After the war, in 1947, he switched to study Biology in Cambridge. At the age of 35, he was still working toward a PhD in Biology, studying the structure of protein using X-ray diffraction. Crick was more of a theoretician than an experimentalist, and a good theoretician he was. He freely criticized other people's good ideas, and filled in the gaps that they had missed.

The instant that Watson met Crick, they found that their intellectual minds clicked. Within half an hour, they were conjecturing the structure of DNA. DNA is made up of nucleotides. Each nucleotide is composed of a phosphate group, a sugar group, and a nitrogen-rich base. However, there are four different bases — adenine, guanine, cytosine and thymine, abbreviated as A, G, C and T. The phosphate group of one nucleotide is linked to the sugar group of another one. Watson and Crick soon decided that they should build a model of the structure of DNA. But which model should they pick to start with? DNA could consist of a long chain of nucleotides, with one linked to the next. Or it could be a closed ring, with one nucleotide joined to the next, until one came back to join the first one. They quickly decided that they should work with the helix model. A helix is a spiral. Mathematically speaking, it is a three-dimensional curve that lies on a cylinder, such that its angle to a plane which is perpendicular to the axis of the cylinder is a constant.

A helix model had recently been proposed by Linus Pauling for the structure of protein. Pauling was a world-famous chemist at Cal Tech, where his associates provided experimental evidence to support his model. Proteins are composed of large numbers of repeating units called peptides, which are attached together to form a large molecule. Thus, protein has a similarity with DNA, which consists of a long chain of nucleotides. Therefore, it would be obvious that Watson and Crick would borrow the helix model, "imitate Linus Pauling and beat him at his own game". They hoped that they could solve the puzzle first, as Pauling was also working on finding the structure of DNA.

The evidence that the structure of DNA was helical could come from X-ray diffraction photos. While the photos that we normally take from a camera are projections of 3-dimensional space on a 2-dimensional plane, diffraction photos are actually projections of an inverse 3-dimensional space on a 2-dimensional plane. Thus, their interpretations are indirect and non-straightforward. One needs to understand how crystallized molecules diffract X-ray in order to explain X-ray diffraction photos. To avoid a false start, Crick invited Maurice Wilkins to come to Cambridge for a weekend so they could see the photos that he had been taking. It did not take any persuasion from them that Wilkins also believed that the DNA structure was a helix. As a matter of fact, Wilkins was showing X-ray diffraction photos of DNA six weeks before Watson arrived at Cambridge, and certain feature in the photos showed compatibility with a helix. However, Wilkins thought that three chains were needed to construct the helix. Also, he doubted that using Pauling's model-building would quickly allow one to determine the structure of DNA. Here, one can see that while some scientists share the same idea, they may deviate for their approaches in solving a problem. The various approaches would just make a difference who will get there first.

Opportunities also quite often play a part in a success story. On October 31, 1951, Sir Lawrence Bragg, director of the Cavendish Laboratory, showed Crick a letter he had just been sent by the crystallographer Vladimir Vand from Glasgow. The letter described a theory for X-ray diffraction by a helix. Vand was hoping that his theory would help to interpret X-ray diffraction photographs of helical molecules.

Crick quickly found an error in Vand's endeavor, and dashed upstairs to consult a physicist, Bill Cochran, who was a young lecturer at the Cavendish. Cochran had independently found faults in Vand's paper, and was wondering what the right answer should be. As a matter of fact, Bragg had been after him for months to derive the helix theory.

That afternoon, Crick went home to nurse a headache. There, he took up the equations again and eventually figured out what the right solution should be. The next morning in the Laboratory, he found that Cochran had also arrived at the same answer, though with a more elegant derivation. A manuscript was written within a few days, and was published the following year in Acta Crystallographica. The authors acknowledged that the same theory was actually derived by Alexander Stokes at King's College a few months earlier. This theory eventually played an important role in enabling Watson and Crick to interpret X-ray data of helixes in the future. We can also see here an example that in scientific research, quite often the same problem would be worked on and solved independently by different researchers at approximately the same time.

10.1.3.3 *Rosalind Franklin at King's College, London*

Rosalind Franklin (1920–1958) graduated with a PhD in physical chemistry from Cambridge University in 1945. After Cambridge, she went to Paris to work on the structures of various forms of coal. She learned the X-ray diffraction technique in Paris, and became quite skillful with it.

In January 1951, she returned to England to work as a research associate in the Medical Research Council's Biophysics Unit at King's College. The Unit was headed by John Randall. As the story goes, Randall wrote Franklin a letter on December 4, 1950, saying that Maurice Wilkins and Alexander Stokes intended to stop working on DNA, and the X-ray diffraction of DNA would be all hers. According to Wilkins, he was not aware of the letter until after Franklin's death.

In July 1951, Wilkins showed some X-ray pictures of DNA at a colloquium at Cambridge. He conjectured that DNA had a universal structure that included a helix. Crick was there. But he barely remembered what Wilkins talked about, as he was not particularly interested in DNA then, not until Watson arrived at Cambridge in the fall of 1951.

Right after Wilkins' talk, Wilkins was confronted by Rosalind Franklin, who simply told him to stop working on DNA, as that work was now hers because Randall said so. Strangely, Wilkins thought that Franklin had been recruited to assist him. To settle the issue, he eventually agreed to hand over to her the DNA crystals that he had been working with, and concentrated his work on other DNA, which he later found did not crystallize.

Throughout the summer, Franklin rebuilt the X-ray apparatus with the assistance of Raymond Gosling, a PhD student that she had inherited from Wilkins. She was then able to take X-ray pictures with the crystals that Wilkins gave her. However, she did not communicate her results to Wilkins, who would have to find out, like everybody else, from a talk she would give at a colloquium he was helping to organize at King's College on November 21 (Wednesday).

Watson was at the colloquium, as he had asked Wilkins to invite him. Crick did not attend, as he was still not treating DNA as his main interest. In her talk, Franklin showed some X-ray photos of DNA in a moister state. This wet or B form of DNA produced diffraction pattern that showed strong evidence that the structure of DNA was helical. Franklin had succeeded in developing an ingenious method to separate the wet B form from the dry A form, and produced clear interpretable patterns of either the B form or the A form.

10.1.3.4 *The triple helix model*

The following morning after Franklin's talk, Crick was quizzing Watson about the new photos shown by Franklin. However, Watson had not taken any note. To make it worse, he had learned crystallography for less than a month, and did not understand some of the jargons that Franklin said. In particular, he did not recall the water content of the DNA samples in the experiment. Fortunately, he did remember some of the key dimensions. Within a few hours, Crick figured that there would be only a few configurations that could fit both Franklin's experimental data and Cochran-Crick helical theory. They might be able to emulate Pauling and build a model with the available information.

For the next few days, Watson and Crick assembled the various atomic models, and finally finished building a three-strand model — a triple helix — on November 26 (Monday). The model was incorrect on several aspects. First of all, the number of backbones was not three. Three had been chosen as it was more consistent with the calculated density of DNA. Dimensions of the molecule could be determined from X-ray diffraction photos. If the weight was measured, the density could be calculated, and the number of strands deduced. Unfortunately, the calculated density eventually turned out to be incorrect, and therefore, the number of strands was not three.

Secondly, the bases were incorrectly put outside the sugar-phosphate backbones that formed the strands of the helix, as Watson and Crick did not know how to fit the bases of different sizes inside the rigid backbones. Thirdly, the phosphates at the backbones were negatively charged, and would repel each other, making it impossible for the three strands of the triple helix to be held together. To overcome this problem, Watson made the wild assumption that positively charged magnesium ions laid inside to hold the strands together. However, there had been no evidence of magnesium in DNA.

On November 27 (Tuesday), Crick phoned Wilkins, telling him that they had come up with a model of DNA, and invited him to check it over. Wilkins travelled up from London the next morning, together with his collaborator Willy Seeds, as well as Franklin and Gosling. After they were shown the model, Franklin proclaimed that it could not be correct. In particular, Watson's recollection of the water content of her DNA samples was wrong. The water content should have been ten times more. Furthermore, Franklin presented evidence that the backbones should be on the outside of the structure. Thus, the model that Crick and Watson built turned out to be a failure.

Some time in December, Wilkins wrote Crick, politely asking him and Watson to stop working on DNA. Later on, Bragg would have come to an agreement with Randall, and told Crick and Watson not to trespass on other people's turf.

10.1.3.5 *The double helix model*

Watson and Crick obliged, but it did not stop them from thinking about DNA. In the last week of May, 1952, Erwin Chargaff, visited Cambridge. Chargaff was a world expert on DNA, and he had discovered an interesting fact about its nitrogen-rich bases — where the number of adenine (A) molecules was about the same as the number of thymine (T) molecules, while the number of guanine (G) molecules was about the same as the number of cytosine (C) molecules. Watson and Crick met Chargaff, and Crick heard of the Chargaff's base ratio for the first time. To Crick, it was valuable information, as he had been trying to figure out how the base pairs would pair up if they were at the inside of the helix.

On January 28, 1953, Watson and Crick saw an advance copy of a paper written by Linus Pauling and his associate Robert Corey, describing a proposed structure for DNA. They quickly found out that Pauling presented a three-chain helix with the phosphates on the inside. The model was somewhat similar to their failed debacle about a year ago. Even more to their surprise, the phosphate groups in Linus's model were not ionized and therefore had no net charge, and his DNA structure was thus not an acid at all.

Two days later, Watson visited King's College and showed Franklin Pauling's manuscript. Franklin asserted that, judging from her latest X-ray data, there was no evidence that DNA was a helical structure. Later on, Watson met Wilkins, who showed him some new X-ray photos, including a photo of the B structure taken by Franklin the previous May. Watson quickly realized that this last photo could only arise from a helical form. Wilkins agreed. However, while Wilkins thought that the model would be three-chain, Watson thought that it might be two-chain, as he thought that significant biological objects come in pairs. Furthermore, the number of strands depended upon the water content of the DNA samples, a value that the King's group admitted might be in error.

Fearing that Pauling would soon find out his own mistake and make another dash for it, Watson approached Bragg to ask for permission to have another go in building the model of DNA. To his relief, Bragg encouraged him to do so.

After some of the pieces were built by the machine shop, Watson spent two days trying to build a two-chain model with the backbones inside and the bases outside. However, he could not build one without violating the laws of chemistry. He then switched to building a backbone-outside model.

On February 8 (Sunday), Wilkins came up to Cambridge for a social visit. During lunch, Watson and Crick tried to persuade Wilkins to start building models. However, Wilkins said that he wanted to put off model-building until Franklin left for Birkbeck College in March. Crick took the opportunity to ask whether they could go ahead and gave it a try. Wilkins reluctantly agreed. Nevertheless, even if his answer was no, the model-building would have gone ahead anyway.

A few days later, Max Perutz, a senior researcher at the Cavendish, showed Watson and Crick a short report written by the King's group for the Medical Research Council (MRC) the previous December. Perutz was a member of a committee appointed by the MRC to evaluate the progress of research carried out at King's. Much debate would be carried on later whether Perutz should show the report to others, but he maintained that it was not marked confidential. In the report, Franklin talked about the shape of the unit cell of the molecule. That information allowed Crick to figure out that the two chains were antiparallel. Now, if the chains were running in opposite directions, the structure would repeat itself after a whole turn of each helix. This vital piece of knowledge would only leave one more piece of puzzle to be figured out. How would the bases fit into the middle of the structure?

Watson had begun to realize the possibility that bases could form regular hydrogen bonds with each other. He first tried the idea of like-with-like pairing, i.e., adenine of one chain paired with adenine on the other, etc. This will form the rung of the staircase. He was soon told by the visiting American crystallographer Jerry Donohue that the idea would not work. Donohue told Watson that the tautomeric forms of the bases that he took out of a textbook were incorrect. As a matter of fact, all textbook pictures were wrong. The enol form should be replaced with the keto form.

Not willing to wait for the new metal plates for the keto bases to be made by the machine shop, Watson cut out cardboard ones instead. On February 28 (Saturday), he came to work early and started playing with the cardboard bases, shifting them in and out of various pairing possibilities. He suddenly realized that "an adenine-thymine (A-T) pair held together by two hydrogen bonds was identical in shape to a guanine-cytosine (G-C) pair held together by at least two hydogen bonds". This discovery formed the last piece of the puzzle of the structure of DNA. As soon as the machine shop had produced the metal plates, Watson and then Crick assembled and double-checked the model until they felt happy with it.

On March 7 (Saturday), Wilkins, not knowing the progress made at Cambridge, wrote Crick a letter, saying that Franklin would be leaving the next week, and they would start building a model. The same day, he was notified that a model had already been built. On March 12, 1953, he came to see the assembled structure. It did not take him long to realize that it had to be correct. He refused Crick's offer to be a co-author of a letter that would be sent to *Nature*. But years later, he said he regretted.

In 1962, Watson, Crick and Wilkins were awarded the Nobel Prize in medicine for their discoveries concerning the molecular structure of DNA. Franklin was not chosen, as, unfortunately, she died in 1958 at the early age of thirty-seven. Nobel Prizes are not awarded posthumously.

10.1.4 *Creative thinking and Ordinary thinking*

The basic cognitive activities in the creative achievement described above can be explained in most parts by the mental processes behind ordinary thinking, which, as mentioned before, consists of memory, planning, judgment and decision.

Memory definitely plays an important role. The knowledge from the participants as well as their colleagues is critical in solving the double helix problem. Both Watson and Crick brought their own expertise to the table. Together with information provided by Wilkins, Frankin, and Donohue, they eventually were successful in assembling the model.

Planning needs to be done to determine what approaches should be taken. Very early on, Watson and Crick would pick the childlike toy construction model building path, while Wilkins and Franklin would tread carefully and think that more experimental data should be gathered. The path would establish who would get there first.

Judgment and decision then follow. Should a two-strand or three-strand model be built? Should the bases be put inside or outside? Current information would be assessed and judged. Watson and Crick first built a three-stand model that was shown to be inconsistent with experimental data. More than a year later, they changed to building a two-strand model. Again, they originally put the bases on the outside, as they did not know how to fit them inside. Only much later on would they learn how the bases could be fit inside.

As one can see, for this particular case study of creativity, almost the whole thinking procedure can be explained in terms of cognitive components in ordinary thinking. The only event that needs clarification in the future is the illumination that Watson felt when he suddenly realized that the A-T pair was identical in shape to a G-C pair.

10.2 Scientific Research and Scientific Method

The discovery of the double helix model also provides an example of how scientific research is being done, and how scientists employ the scientific method of observation, hypothesis, and experiment.

Observation implies gathering and filtering knowledge — knowledge that one has learned and stored in one's brain and needed to be sorted out for their relevance to the current problem, knowledge that one requires to gather from future experimental data, and knowledge collected by discussing with colleagues and other scientists. This just shows why collaboration is important. While Watson and Crick were working toward the same goal, the bickering between Wilkins and Franklin would mean the slowing of the entire research process. Furthermore, discussion with others definitely helps. This can be exemplified with the conversation of Watson with Donahue, who pointed out the mistake of the tautomeric form of the bases as depicted in all the standard textbooks.

Hypothesizing plays an important role in scientific research. It jump-starts the whole discovery process. Is DNA, rather than protein, the material carrying the genetic information? Is its structure a helix? If so, how many strands would it have? Should the bases stay inside or outside? The results of the hypotheses should be consistent with existing experimental data, as well as confirmed by future experimental results.

This is why experiment is significant for demonstrating that whatever is hypothesized is not pure conjecture, but does represent reality. The earlier triple helix model as proposed by Watson and Crick contained ten times less water than the experimental finding of Franklin, and could not be correct. They had to go back to the drawing board and try again. Needless to say, it is unavoidable that mistakes may be made when one hypothesizes. One just has to try again and get it right. That is what Watson and Crick did, and eventually came up with the double helix model. The model withstood the test of experimental data and therefore considered valid.

10.3 Can we be more creative?

The answer is: most definitely, yes. One does not need to be a genius to be creative. As it is shown above, creative thinking process is no different than the ordinary thinking process. So, one does not need any superbrain power to be creative. We can be more imaginative if we follow and practice the scientific method of observation, hypothesis and experiment.

Observe our surroundings, and try to notice the relationship among various objects or concepts. Read the newspapers. Talk to other people. We may pick up information that we are particularly looking for, or even information that we are uncompletely unaware of.

Hypothesize why certain events happen the way they do. Come up with ideas how problem situations can be handled. The earlier we can come up with a hypothesis, the quicker we can act on the issue. Hypothesizing gives us a sense of direction, and allows us to channel our energy toward a certain goal.

Nevertheless, a hypothesis is only an idea, and it needs to be tested to see whether it actually works. That is why experiments have to be performed. The proof of the pudding is in the eating. If one idea does not work out, try another one until we get it right.

Some basic knowledge comes in handy if we want to be creative. It would help if we learn some fundamental concepts in physics, chemistry and biology. In addition, learning some elementary mathematics will definitely help, as we will show in the next chapter.

Chapter 11

Mathematics

Mathematics, even some simple arithmetic, is so important in solving some of the everyday problems, that we think a whole chapter should be written on it.

Let us take a look at an example. When we see an advertisement which says "Buy one, get the second one at half price", we should be able to figure out what exactly does it mean, and how much discount are we actually getting. Is it a better deal than another company that advertises 30% off?

The answer is no. "Buy one, get the second one at half price" simply means a 25% discount, i.e., if you buy two items. And it means that you are almost coerced to buy two items. You can, of course, buy one item, but then you will not get any discount.

If the company advertises "Buy one, get the second one (with equal or lesser value) at half price", that means that the maximum discount that you are getting is 25% off.

Now, you may want to test this question on your friends. If a company has a 100% mark-up of a merchandize from its cost price, and now it is advertising 50% off, is the company making any profit on that item? You may be surprised how many people did not get it right.

If the cost price of the merchandize is $1, a 100% mark-up means that the selling price is set at $2. Now if the company is advertising a 50% off, that means it is selling the item at a sale price of $1. In that sense, the company is actually not making any profit on that item. Now, let us take a look at another question.

Let's say, when we go shopping, we have to pay a government sales tax of 15%. Let us also assume that we have a 10% discount coupon on a merchandise. Does it make any difference to us if the discount coupon is applied before sales tax or after sales tax? Again, you may be amazed how many people got it wrong.

A sales tax of 15% (= 0.15) means that the total payment would equal to the price times 1.15 (= 1 + 0.15). A discount of 10% (= 0.10) means multiplying the price by 0.90 (= 1 – 0.10). Whether one multiplies the price by 1.15 and then 0.90 or by 0.90 and then 1.15 would not have make any difference. Thus, the amount that the customer pays would be the same whether the discount coupon is applied before or after.

Now, let us take a look at the following interesting example.

Example 1 Buy one, get one free

"Buy one, get one free" means a 50% discount. (However, it is not exactly equivalent to 50% discount, as you wind up having two items instead of possibly one.)

While Lucy was looking at an advertising flyer of Company A, she noticed that a certain item was advertised as "Buy one, get one free". By chance, the same item, shown in the flyer of Company B, was also on sale at a discount of 40%. She also noted that the "50% discount" sale price at Company A was larger than the 40% discount sale price at Company B. This would imply that the original price of the item in Company A was larger than that in Company B.

Lucy quickly came up with the hypothesis that the markup at Company A must be larger than 100%. Furthermore, Company A must be a more expensive store to shop at than Company B. She later compared the prices of several items in both companies, and found out that the prices of Company A were, in general, 10–15% higher than those of Company B. Obviously, from then on, she would shop at Company B.

The following example will also show us that knowing some mathematics will help us perform a simple cost-benefit analysis.

Example 2 Birthday cake

It was the daughter's twelve-year-old birthday. The family of four drove to an ice cream parlor to purchase an ice cream birthday cake, as that was what the daughter wanted. When they got to the store, the father had to go use the washroom. By the time he came back, the children had already bought the cake. They had bought an ice cream cake 8" in diameter.

The father then looked at the price list that was posted on the wall. While ice cream cake with an 8" diameter was selling for $20.00, an ice cream cake with a 10" diameter was selling only for $22.00. So he asked the children why did they not get the 10" cake. They told him that they probably could not eat that much.

But the father said that ice cream was a nonperishable item, and they could have kept the remaining cake in the freezer.

The father then asked the children whether they knew what was the formula for the area of a circle. They hesitated, and the father explained. The area of a circle is equal to πr^2, where r is the radius, i.e., the area is proportional to the square of the radius. As radius is equal to half of the diameter, the area of a circle is proportional to the square of the diameter. Assuming the 10" diameter cake has the same height as the 8" diameter cake (that was indeed the case), the volume of the cake will be proportional to the square of the diameter. That simply means that the 10" cake is larger in volume than an 8" cake by a factor of $(10/8)^2 = 1.5625$, i.e., the volume of the 10" cake is approximately 56% more than that of the 8" cake. However, the price of the 10" cake was only $(22 - 20)/20 \times 100\% = 10\%$ more. By paying 10% more money, they would have got an ice cream cake that would be more than 50% more in volume or weight. In that sense, they should have bought the 10" cake.

The children agreed. They realized that they had just learned an important mathematics lesson from their father.

Now, we will take a look at an example where a little mathematics would make a lot of difference — a difference of about $40,000.

Example 3 Buying an apartment

Dr. McGrath and his wife have two daughters of one year apart, Justine and Sarah. He is a cardiologist in a hospital in Cornwall, Canada, and his wife is a stay-at-home mom. Since he is well respected and well paid, he obviously would like his daughters to follow his footstep.

In the year 2005, Justine received her Bachelor degree, and was admitted to medical school at University of Toronto. The parents were happy. The next year, Sarah was also accepted to the same medical school. The parents were overjoyed, and threw a party to celebrate both their daughters going to medical school.

At the party, Dr. McGrath was saying that their daughters would be living together and they would be looking for an apartment to rent. A friend of his, Michael, overheard the conversation, and asked him why did he not simply buy an apartment in Toronto for them to live in. Dr. McGrath thought that was not a bad idea, and he said that he would loan the girls the down-payment so that the apartment would be in their names, and the girls would pay the mortgage interest which would probably be cheaper than if they had to pay rent. When they sold the apartment in the future, they could pay him back the loan, and they could pocket any profit, as the profit from selling a principal residence (where the homeowner lives) in Canada is tax-free.

Michael then suggested to Dr. McGrath that he should buy the apartment as investment himself, and rent it back to his daughters. He could then write off any losses incurred during the year. He could sell the apartment in ten years when he retired, as his marginal income tax rate would be lower than the rate he was paying now. Furthermore, the profit would be considered as capital gain, where only 50% is taxable according to the Canadian tax regulation. In any case, the whole family would be better off if the apartment is under his name than if the apartment were in his daughters' name.

Dr. McGrath was not convinced. So Michael took out a piece of paper, and did the mathematics as follow:

Price of apartment = \$250,000

Assume they put nothing down. That is, they borrow 100% of the price of the apartment from the bank. This can be achieved by borrowing 25% from their home equity line-of-credit, and 75% as mortgage for the apartment.

Mortgage rate = 6% = 0.06

Mortgage interest = 0.06 × \$250,000/year = \$15,000/year = \$1250/month

Condo fee (paid to maintain common areas of the whole apartment complex) = \$250/month

Propery tax = \$3,000/year = \$250/month

Utilities (e.g., electricity) = \$300/month

Maintenance, insurance, etc = \$100/month

Therefore, total expenditure/month = \$1250 + \$250 + \$250 + \$300 + \$100 = \$2,150

Scenario 1: Apartment under daughters' names (Daughters as homeowners)

Assume 5.5% increase in house price per year, and house is sold 10 years later.

Use simple rate of return and not compound rate of return for the calculation.

In 10 years, house price would increase by 55% (= 0.55), which is the net gain as the house is a principal residence.

Scenario 2: Apartment under Dr. McGrath's name (Dr. McGrath as investor)

Rent paid to him by his daughters = \$1,000/month

Total loss per month before tax for Dr. McGrath = \$2,150 – \$1,000 = \$1,150

Marginal income tax rate of Dr. McGrath = 0.4

Total loss per month after tax for Dr. McGrath = \$1,150 × (1 – 0.4) = \$690

(This is equivalent to a tax recovery per month of \$1,150 × 0.4 = \$460)

Therefore, % gain per year (as compared to house owned by his two daughters) = \$460 × 12/\$250,000 × 100% = 2.208%

Total % gain for 10 years (as compared to house owned by his two daughters) = 10 × 2.208% = 22.08%

Assume house is sold 10 years later when marginal income tax rate of Dr. McGrath is 0.25

According to the Canada Tax Law, 50% of the profit (capital gain) is taxable.

Therefore, net gain by Dr. McGrath for selling the house = 55%/2 + 55%/2 × (1 – 0.25) = 0.48125

Total net gain for Dr. McGrath = 0.48125 + 0.2208 = 0.70205

which is greater than 0.55 in Scenario 1.

For an apartment price of \$250,000, the difference of percentage for Scenario 2 over Scenario 1 would mean an actual dollar amount of $(0.70205 - 0.55) \times \$250,000 = \$38,012.50$

Thus, if the apartment is under Dr. McGrath's name instead of under his daughters', the family, as a whole will gain \$38,012.50 more in 10 years' time.

Had Dr. McGrath paid for his loss every year (and he definitely can with his high salary), he will pocket after tax, at the end of 10 years after the house is sold $0.48125 \times \$250,000 = \$120,312.50$

Michael showed his calculation to Dr. McGrath, who was finally convinced. Six months later, Dr. McGrath bought an apartment within walking distance to the University of Toronto campus. The apartment was put under his name.

This example just shows that a little bit of mathematics can go a long way.

Example 4 Currency exchange

People these days travel much more often than they used to. When we travel to a different country, we need to use the currency of that country. How do we know whether a bank or a foreign exchange firm is giving us a good exchange rate? There is an easy way to find out. Simply ask them what are their sell rate and buy rate. Sell rate is the rate that they sell to us, i.e., we buy from them. Buy rate is the rate that they buy from us, i.e., we sell to them. Subtract the buy rate from the sell rate. Divide the difference by the buy rate or the sell rate, and then multiply the result by 100%. The equation is shown as follows:

approx. diff % = (sell rate – buy rate)/(sell rate or buy rate) × 100 (1)

A more accurate result is shown in Eq. (2) as follows:

diff % = (sell rate – buy rate)/((sell rate + buy rate)/2) × 100% (2)

The result calculated from Eq. (1) is approximately equal to that of Eq. (2). For all practical purpose, Eq. (1) should suffice. However, just for the sake of argument, we will use Eq. (2) in the following discussion.

If the diff % is less than 3%, then the exchange rate we are getting would be quite reasonable. If it is higher than 3%, then the exchange rate would be considered to be on the high side.

Let us take a look at an example of the exchange rate between Canadian dollar and Euro.

At a certain time on March 28, 2007, a Canadian bank was selling 1 Euro cash at 1.6021 Canadian dollars, and buying 1 Euro cash at 1.4954 Canadian dollars. It is also selling 1 Euro traveler's cheque at 1.5821 Canadian dollars, and buying 1 Euro traveler's cheque at 1.5039 Canadian dollars. These sell and buy rates are listed in Table 1.

Table 1. Exchange rate of bank for selling and buying Euro cash and traveler's cheque with Canadian dollars.

Bank	Cash	Traveler's cheque
Sell	1.6021	1.5821
Buy	1.4954	1.5039
average	1.5488	1.5430
diff %	6.88	5.06
½ diff %	3.44	2.53

In Table 1 above, average is defined as:

$$\text{average} = (\text{sell rate} + \text{buy rate})/2 \quad (3)$$

The averages calculated in Table 1 are approximately equal to the market exchange rate, which is about 1.5435 at that time on that date. The market exchange rate is the rate that is traded in the financial market, and it varies throughout the day. This rate can be found in the Internet. The bank buy and sell rate also varies with the market exchange rate. (It should be noted from Table 1 that while the average of the sell rate and the buy rate for traveler's cheque is just about the same as that of the market exchange rate, the average of the sell rate and buy rate for cash is larger than the market exchange rate by about 0.3%. We will come back to this point later)

As the diff % in Table 1 is larger than 3% for both cash and traveler's cheque, the exchange rate of the bank is somewhat on the high side.

½ diff % is half of the diff %. 3.44% calculated in Table 1 is the one way fee that we are losing when we buy Euro cash with Canadian money, or if we sell Euro cash back to the bank to get back Canadian money. If we buy Euro cash from the bank with Canadian money, and then sell the Euro cash back to the bank right away, then we will be losing 2 × 3.44% = 6.88%. That is, for every $100, we will be losing $6.88

Of course, we can purchase traveler's cheque from the bank, and we can get a slightly better exchange rate. The cheque rate is always cheaper than the cash rate. The reason is that the bank does not have to hold the actual cash, as it is costly for the bank to ship out the cash when a large amount has been collected.

2.53% is the one way fee that we are losing when we buy Euro traveler's cheque with Canadian money, or if we sell Euro traveler's cheque back to the bank to get back Canadian money. If we buy Euro traveler's cheque from the bank with Canadian money, and then sell the Euro traveler's cheque back to the bank right away, then we will be losing $2 \times 2.53\% = 5.06\%$. That is, for every $100, we will be losing $5.06.

In addition, the bank charges 1% issuing fee for traveler's cheque. So, if we are purchasing Euro traveler's cheque from the bank, we are actually losing 3.53 %. This is larger than the 2.5% fee that credit card companies usually charge for any foreign currency transactions (the current market exchange rate is used in the transactions by the credit card companies). Therefore, when we are travelling in Europe, using our credit card can save us a bit of money than using Euro traveler's cheque that we can purchase from the bank.

One can actually save that 1% issuing fee for traveler's cheque, as one can purchase traveler's cheques from a Canadian travel association, which offers no issuing fee for members and "competitive exchange rate" as they claim on their website. However, upon checking their exchange rate, it was found that they were selling Euro traveler's cheque at 1.6214 Canadian dollars on March 28, 2007, and that is a whooping 5.05% more than the market exchange rate. So, this % is much larger than the 3.53% that will cost us if we purchase traveler's cheque from the bank. Thus, unless one is observant enough to compare their rate with those of the banks or other institutions, one may be tempted to buy from this Canadian travel association, as they do not have issuing fee.

The best exchange rate one can get seems to come from one foreign exchange firm, which guarantees the best rates on cash on its website. Its sell and buy rates are listed in Table 2. However, even though it lists its sell rate for cheque, it does not issue traveler's cheque. (It sells cheque that has to be deposited in a bank account in a foreign country). But they do buy traveler's cheque from the customers.

Table 2. Exchange rate of a foreign exchange firm for selling and buying Euro cash and cheque with Canadian dollars.

Foreign exchange	Cash	Cheque
Sell	1.5642	1.5625
Buy	1.5218	1.5234
average	1.5430	1.5430
diff %	2.78	2.53
½ diff %	1.39	1.27

As the diff % in Table 2 is smaller than 3% for both cash and cheque, the exchange rate of the foreign exchange firm is quite reasonable.

1.39% is the one way fee that we are losing when we buy Euro cash with Canadian money, or if we sell Euro cash back to them to get back Canadian money. If we buy Euro cash from the foreign exchange firm with Canadian money, and then sell the Euro cash back to them right away, then we will be losing 2 × 1.39% = 2.78%. That is, for every $100, we will be losing $2.78.

The averages calculated in Table 2, are approximately equal to the market exchange rate, which is about 1.5435 at that time on that date.

An interesting observation for the bank averages is: while the average of the sell rate and the buy rate for traveler's cheque is approximately the same as that of the market exchange rate, the average of the sell rate and buy rate for cash is always larger than the average of the sell rate and buy rate for traveler's cheque by about 0.3%. An example can be seen in Table 1, where 1.5488 is larger than 1.5430 by 0.37%. This, of course, is to the advantage of the bank, as they definitely sell more cash than buy cash back. And it also means that, if we are buying Euro cash from them, instead of losing only 3.44% (as shown in Table 1), we are actually losing (3.44% + 0.37%) = 3.81%.

So, it does make quite some difference where we do our foreign exchange before we go travelling. Let us say that we are going to spend Canadian $10,000 in Europe on a vacation. If we purchase Euro traveler's cheque from the Canadian travel association, we will be losing 5.05%, which would be equivalent to $505. If we purchase Euro cash from the bank, we will be losing 3.81%, i.e., $381. However, if we purchase Euro cash from the foreign exchange firm, we will be losing only 1.39%, i.e., $139. And, if we are not comfortable in carrying so much cash in our pocket, then we can use our credit card, and we lose only $250.

Another way is to purchase Canadian dollar (or local currency of your country) traveler's cheque, and sell it at a foreign exchange firm in the foreign country that you are visiting. Banks sometimes do not charge traveler's cheque issuing fee for their good customers. Furthermore, the Canadian travel association also does not charge traveler's cheque issuing fee for its members. If so, for $10,000 expenditure, you are probably losing only about 1.27%, i.e., $127 (using Table 2 as a guide), providing you can find in that foreign country a foreign exchange firm that can provide a good exchange rate. (Usually, if you join a tour, the tourist guide can show you where to go for a foreign exchange firm that has a good rate.)

Again, this example shows that a little mathematics can save us quite some money.

Example 5 Investment

Trisha lives in Winnipeg, Canada. In February, 2008, she attended a financial seminar organized by an investment company, and eventually arranged to meet with Sam, a financial consultant from that company.

Sam started off telling her that one only needed to pay off the mortgage of the house before retirement. Many people tried to pay off their mortgage as quickly as possible. But that was a mistake. As the mortgage rate was always low, one should use the money to purchase mutual fund, which would yield a good return. To give her an example, Sam pointed to a chart where the S&P (Standard and Poor) Index was plotted. If someone invested $10,000 in the stocks of the S&P Index in the beginning of 1996, she would get $37,800 by the end of 2007.

Trisha then asked Sam what was the average rate of return for those twelve years. Sam was somewhat dumbfounded, as he did not know the answer. Nor did he know how to calculate the rate of return from his financial calculator, or whether his calculator had such built-in software to begin with. All he knew was that he could calculate the future value from a present value, but only if the rate of return was given. Trisha then took a piece of paper and her scientific calculator, and worked out the calculation as follows:

Let r be the average rate of return for those twelve years.

Assuming a compound rate of return,

$$10{,}000\,(1 + r)^{12} = 37{,}800$$

In the above equation, 10,000 is the present value, and 37,800 is the future value 12 years later. Simplifying and taking natural logarithm on both sides of the equation, one gets

$$12\ ln\,(1 + r) = ln\,3.78$$

$$(1 + r) = \exp\,(\,ln\,(1 + r)) = \exp\,(\,(ln\,3.78)\,/12\,) \approx 1.12$$

where ln is the natural logarithm, and exp is the exponential.

The average rate of return, r, is therefore equals to 0.12 = 12%. As the management fee for a mutual fund is approximately 2%. The net average rate of return is about 10% before tax.

The mortgage rate at the end of February 2008 is 7.25% for a one-year term, and 7.29% for a five-year term. Thus, the financial consultant's suggestion that one should invest in mutual fund instead of paying off the mortgage does have some truth in it. This, of course, will depend on whether a particular mutual fund will perform as well or even outperform the market index. However, if the after-tax mutual fund gain is lower than the mortgage rate, then one should pay off one's house instead of investing in the mutual fund.

Trisha showed Sam her calculation. Interestingly, Sam asked Trisha whether he could keep that piece of paper where she did the calculation. He actually learnt something from a potential client.

Example 6 Average return rate of an investment with regular contribution

There is a certain reason why investors are interested in the return rate of an investment — as the return rate can be compared to the current inflation rate, interest rate and mortgage rate.

In general, interest rate of a term deposit in the bank is less than the inflation rate, which simply means that if one leaves the money in the bank, the money would not keep up with inflation.

Mortgage rate is larger than the interest rate, as the bank has to make money for loaning money out.

In general, investment in stocks and mutual funds would beat the inflation rate. However, the stock market is usually volatile, and any such investment should be long term. A number of investors would buy into a stock or mutual fund using periodic contribution, so that they sometimes buy at a low price, and sometimes they buy at a high price, depending on the value of the stock or mutual fund at that moment. But then, the question is: what is the average rate of return of their investment? It does not seem as if too many people know how to calculate it.

Trisha remembered that, in January 3, 2007, she asked her financial advisor at the bank to automatically withdraw $300 from her bank account every first day of the month to purchase a certain mutual fund. She wanted to set up this mutual fund account for her retirement. On February 2, 2008, she noted that there was $4,019 in her mutual fund account. So, up to then, she had paid thirteen payments of $300, i.e., a total of $3,900. Now she was wondering what the annual average return rate of her mutual fund was.

So she asked her financial advisor at the bank whether he knew how to do the computation. He told her that quite a number of his clients had asked the same question, but he did not know the answer. And that piqued her interest.

Trisha figured that, to a good approximation, she could consider her mutual fund investment to be a simple annuity. An annuity is a type of investment where fixed amounts are deposited or paid at regular intervals over a set period of time. The annuity is called a simple annuity if the payment interval corresponds to the interest conversion period. For example, if the interest conversion period is a month, then the payment interval is a month.

Trisha then asked her accountant boyfriend whether he knew how to calculate the interest rate or return rate of a simple annuity. However, her boyfriend did not have any clue how the rate could be calculated. So he asked his accountant friends. One of them said that one could look it up in a financial table. There are tables that list the future value of a simple annuity, with the amount of regular payment, number of payments and interest rate. Another friend said that he believed there was a program in the financial calculator to do such a computation.

Trisha did not have any financial tables, nor financial calculators. In any case, a financial table will only yield an indirect way to find the interest rate, and will give only an approximate answer. So, she tried to calculate it herself. She looked up her high school mathematics book, and found the chapter on simple annuity.

The formula for calculating the accumulated or future value of a simple annuity, F, is given as

$$F = r\,[(1 + x)^n - 1]/x \qquad (4)$$

where r is the regular payment

n is the number of interest conversion periods or the total number of payments

x is the interest rate per conversion period

Unfortunately, the above equation would not allow her to write out the interest rate (or the return rate, in her case), x, explicitly in an analytical expression. The interest rate has to be estimated graphically or calculated using numerical analysis by applying, e.g., Newton's method of root-finding. She chose the simpler method of estimating it graphically.

Re-writing Eq. (4), she got:

$$F\,x = r\,[(1+x)^n - 1] \quad (5)$$

The left-hand-side of the equation, when plotted against x, will yield a straight line graph. The right-hand-side of the equation, when plotted against x, will yield a curve. The point where the straight line meets the curve (other than the origin of the graph) will yield the answer for x.

Employing $F = \$4{,}019$, $r = \$300$, $n = 13$ months and x being the return rate per month, Trisha calculated the table as follows by using Microsoft Excel:

Table 3. Estimation of the interest (or return) rate of a simple annuity.

x	$F\,x$	$r\,((1+x)^n - 1)$	$F\,x - r\,((1+x)^n - 1)$
−0.002	−8.04	−7.71	−0.33
−0.001	−4.02	−3.88	−0.14
0	0	0	0
0.001	4.02	3.92	0.10
0.002	8.04	7.89	0.14
0.003	12.06	11.91	0.14
0.004	16.08	15.98	0.10
0.005	20.095	20.09586	−0.00086
0.006	24.11	24.26	−0.15

In Table 3, the left-hand-side and the right-hand-side of Eq. (5) is calculated in column 2 and column 3 respectively with x as a variable. When the numbers in the two columns equal to each other (other than when $x = 0$), x is determined. From the table, it can be seen that the estimated monthly return rate, x, is approximately equal to 0.005, thus yielding an annual average return rate of $12(0.005) = 0.06 = 6\%$.

The answer x can be shown more clearly by subtracting column 3 from column 2, as shown in column 4, which shows their difference. We will define u to be the difference:

$$u = F\,x - r\,[(1 + x)^n - 1] \tag{6}$$

Column 4 is plotted against the monthly return rate x in the following figure. The return rate is determined when the difference curve crosses the x-axis (other than the origin, i.e., $x = 0$), the x-axis being the horizontal axis. When $x = 0$ in Eq. (6), $u = 0$, which explains why the u curve cuts the origin. That point simply means that the future value is equal to the sum of all the regular contributions when the interest rate, x, is equal to zero. The other root of Eq. (6) is the solution that one is looking for.

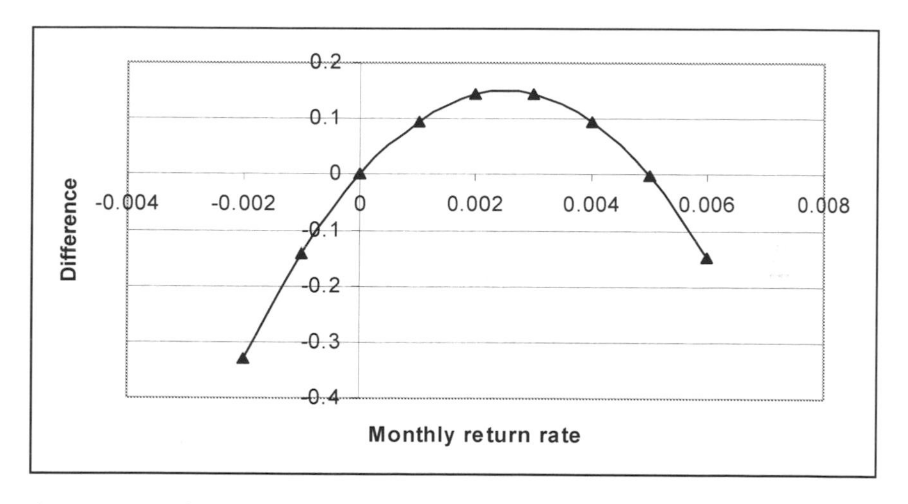

Figure 1. Estimation of the interest (or return) rate of a simple annuity. The point where the curve cuts the x-axis (other than the origin where $x = 0$) yields the interest (or return) rate.

The figure shows that the monthly return rate is 0.005, which yields an annual average return rate of 6%. Trisha was reasonably happy with 6%, considering that the market had dropped quite a bit lately.

The whole calculation and plotting had taken her only about five minutes of typing and programming in Microsoft Excel. It just shows that knowing some mathematics proves to be quite handy sometimes.

Example 7 Average return rate of an investment with initial and regular contributions

Trisha later told her girlfriend Melanie that one could easily find out graphically the average return rate of an investment with a regular contribution. Melanie then asked her whether she could modify the expression to include an initial contribution as well. She had put in an initial contribution of $1,000 in a mutual fund on February 1, 2006, and then starting on March 1, 2006, she regularly contributed $250 on the first day of the month. She checked her investment on April 2, 2008, and there was $8,061 in it. The number of regular payment was thus twenty-six. She wondered what was the average rate of return.

Trisha said that modifying the expression would not be a problem. Modifying Eq. (4) to include an initial contribution, the future value, F, is given as

$$F = P\,(1 + x)^n + r\,[(1 + x)^n - 1]/x \tag{7}$$

where P is the initial payment

r is the regular payment

n is the number of interest conversion periods or the total number of payments

x is the interest rate or the average return rate per conversion period

Re-writing Eq. (7), we get:

$$F\,x = r\,((1+x)^n - 1) + xP\,(1+x)^n \tag{8}$$

Define

$$u = F\,x - r\,[(1+x)^n - 1] - xP\,(1+x)^n \tag{9}$$

By plotting u versus x, the average return rate can be determined from where the curve crosses the x-axis (other than the origin), the x-axis being the horizontal axis. This will take only a few minutes of typing and programming in Microsoft Excel.

Example 8 Pension

The Canada Pension Plan (CPP) requires all Canadians over the age of eighteen to contribute a certain portion of their earnings income to a nationally managed pension plan. A Canadian can then apply for the CPP retirement pension at age 60 or after.

Janet retired at age 58 in the year 2006. A year and a half later, she received a letter from Service Canada, saying that she could receive a retirement pension from CPP starting at age 60. The pension would be $700 per month. However, she could also choose to start receiving her pension at age 65. If so, the pension would be $1,000 per month. That is, if she choose to start receiving her pension at age 60, the amount would only be 70% (= 0.7) of what she would have got at age 65 had she chosen to start receiving her pension at age 65. As she expected herself to live to age 85, she was wondering which option would be more beneficial to her. (The life expectancy for Canadian women in 2006 is 82.6 years.)

So, she sat down and did some calculation.

Let n be the number of years after she turns 60, where, at $(60 + n)$ years old, she will receive the same total pension amount whether she starts her CPP at age 60 or 65. And let b be the pension amount that she would get in one year if she starts her CPP at age 65. To solve for n, she wrote:

$$0.7\, b\, n = b\, (n - 5) \tag{10}$$

The left-hand-side of Eq. (10) is the total pension amount that she would get for n number of years if she starts collecting CPP at age 60. The right-hand-side of Eq. (10) is the total pension amount that she would get for $(n - 5)$ number of years if she starts collecting CPP at age 65.

Simplifying Eq. (10),

$$0.7\, n = n - 5 \tag{11}$$

Solving for n

$$n \approx 16.7$$

Therefore, when Janet turns (60 + 16.7) = 76.7 years old, she will have received the same total amount of pension irrespective whether she starts her CPP at age 60 or 65. After 76.7 years old, she will gain \$(1000 – 700) = \$300 more per month if she applies her CPP at age 65 instead of 60. As she bet that she would live to 85 years old, she would definitely be better off to wait until she is 65 to start her CPP.

But wait, there may be a cost of living adjustment every year for the CPP. So she called Service Canada to find out, and was told that indeed there was a cost of living adjustment every year once someone started taking the pension. The cost of living adjustment averaged about 2% for the last few years. However, the cost of living adjustment would kick in only after one starts taking the pension. That is, if she starts her pension at age 65, she would still only get \$1,000 per month. Thus, Janet thought she should re-do her calculation to take into account the cost of living adjustment.

There was also one more factor that she would like to take into account. For the money she gets, she can put it in the bank to gain interest. The interest rate thus needs to be included in the calculation.

Let r be the total rate of return per year.

r = cost of living adjustment + interest rate ≈ 0.02 + interest rate

Define

$$s = 1 + r \tag{12}$$

To solve for n, Janet wrote:

$$0.7b(1 + s + s^2 + \ldots. + s^{n-1}) = b(1 + s + s^2 + \ldots. + s^{(n-5)-1}) \tag{13}$$

The left-hand-side of the Eq. (13) is the total cost-of-living adjusted pension amount with interest that she would get for n number of years if she starts collecting CPP at age 60. The right-hand-side of the equation is the total cost-of-living adjusted pension amount with interest that she would get for $(n - 5)$ number of years if she starts collecting CPP at age 65.

By using the summation of a geometric series, Eq. (13) can be reduced to

$$0.7\,((s^n - 1)/(s - 1)) = (s^{n-5} - 1)/(s - 1) \tag{14}$$

and n can be solved to be

$$n = \mathit{ln}\,(\,0.3/(s^{-5} - 0.7\,))/\mathit{ln}\ \mathrm{s} \tag{15}$$

where ln is the natural logarithm.

Using Eq. (13), Janet then wrote down the following table:

Table 4. The amount of interest rate where one will receive the same total cost-of-living adjusted pension amount with interest at (60 + n) years old, whether one starts her CPP at age 60 or 65. n is the number of years after one turns 60. Cost of living adjustment is estimated to be 2%.

Interest rate (%)	n
0	19
1	20.7
2	23
3	26.2
4	31.7
5	46.4
6	No solution

The above table means that, with cost of living adjustment at 2% and interest rate at 4%, Janet will get equal total amount in pension at the age of (60 + 31.7) = 91.7 years old, whether she starts her CPP at age 60 or 65. When interest rate hits 6%, it is definitely a good idea to start CPP at age 60, as the total pension amount will never be able to catch up had one started at age 65.

As Janet expects herself to live to about 85, and interest rate was approximately 4% at that time, she decided to apply for her CPP starting at age 60.

For someone that needs the CPP money at age 60, the scenario can be analyzed as such. If he does not apply for the CPP at age 60, he will have to borrow money from the bank at a rate of more than 7%. Judging from Table 4, he will wind up getting less money in total than if he applies CPP at age 65. Therefore, he should definitely apply for CPP at age 60.

Thus, this calculation shows that the result depends very much on how the problem situation is modeled. Had cost of living adjustment and interest rate not included in the model, Janet should apply for CPP at age 65 in order to get the most out of the CPP. If they are included, she should apply CPP at age 60.

Consequently, we can say that any model needs to fit reality in order for someone to make a reliable decision.

Example 9 Self-storage unit

Alice lives in Rochester, USA. She goes to Hong Kong (HK) to visit her older sister, Jennifer, about once a year. Hong Kong is one of the most expensive places in the world to live in. In the year 2006, the price of one square foot of residential area in Hong Kong was about US $500, as compared to about US $150 in USA or Canada. Nevertheless, Jennifer, in her early sixties, was doing fine. She owned her own home, and had some rental apartments to provide for her retirement.

The net rate of return for renting out an apartment in Hong Kong was approximately 3.5%. Thus, renting out apartments was actually not exactly a good investment, as one could get a 5% interest rate for a term deposit in a bank at that time. However, house price will increase. If one assumes a conservative estimate of an average increase of 3% in house price per year, the total rate of return would be (3.5 + 3) = 6.5%, which made rental apartments not such a bad investment.

In October 2006, Alice flew to Hong Kong to visit Jennifer. During lunch, Jennifer told Alice that she had run out of space to put her old furniture and other household items in her own house and had just rented a self-storage unit in an industrial area for storing them. The self-storage unit she rented was basically a

900 square feet (84 square metre) room with high ceiling. She was paying HK \$2,800 (≈ US \$360) for the rent per month. Alice asked how much was the purchasing cost of a self-storage unit of that size, and was told that it cost approximately HK \$330,000. Alice then asked Jennifer why did she not buy a unit instead of renting one. Alice further explained. After deducting property tax and other expenses, she estimated that the owner of the self-storage unit could net \$2,500 out of the \$2,800 rent. That meant his yearly rate of return was $2{,}500 \times 12/330{,}000 \approx 9\%$, which made renting out self-storage unit a pretty good investment as compared to renting out an apartment. This high rate of return would also imply that it would be worthwhile buying a unit instead of renting it.

Jennifer then told Alice that her husband actually told her that, at their age, they should not be buying any properties. They should be selling some of their rental properties, so that they would have fewer problems to deal with. However, Alice pointed out that owning a self-storage unit for their own use did not contribute as much problem as renting an apartment to tenants. Furthermore, the rent that she would have paid in 11 years would have covered the cost of the self-storage unit, as ($1/9\% = 1/0.09 \approx 11$). Alice expected that Jennifer would live more than a further 11 years. In addition, chances were that her children might need a self-storage unit in the future, and she could leave the unit to her children.

Jennifer agreed, and started looking to purchase a self-storage unit.

As we can see from the above examples, mathematics does contribute to solving some of the everyday problems. Working out the numbers does affect our financial decisions, as well as the cost-benefit analysis of performing certain tasks.

Chapter 12

Probable Value

For a certain problem, we may come up with several plausible solutions. Which path should we take? Each path would only have certain chance or probability of success in resolving the problem. If each path or solution has a different reward, we can define the probable value of each path to be the multiplication of the reward by the probability. We should, most likely, choose the path that has the highest probable value. (The term "probable value" is coined by us. The idea is appropriated from the term "expected value" in Statistics. In this sense, expected value can be considered as the sum of all probable values.)

Looking at a problem situation from different perspectives, we may, for example, come up with three problem definitions, and they may have 2, 4 and 3 solutions respectively. We would then have 2 + 4 + 3 = 9 paths leading to a destination. We would need to estimate the probability of success of each path, and evaluate the probable value thereafter.

Estimating the probability of an event can be interpreted as hypothesizing the chance of the future after making observation of the past and the present. To forecast that a certain incident will happen, we need experience and information to evaluate the surrounding circumstances.

Given two paths, we may not choose the one that has a higher probability of success. Instead, we may want to choose the one with a higher probable value, i.e., it may have a smaller chance of success, but a higher reward (or less effort, or less inconvenience), as the following examples will show.

Example 1 Trip to universities

Heather and George have twins, one boy, and one girl. In the year 2003, they graduated from high school at the same time. The boy was accepted to study Computer Science at Queen's University in Kingston, and the girl was accepted to study Business Administration at the University of Toronto in Toronto.

Heather and George had to drop their children off at the universities before September, when the school term started. They live in Ottawa. Kingston is about halfway between Ottawa and Toronto. For George, the most efficient way was to drive on Highway 401, dropped their son off at Queen's, had lunch, and then went on to Toronto after. It would take about two hours to drive from Ottawa to Kingston and then another two hours to drive from Kingston to Toronto. They had a minivan. He believed that, in all probabilities, he could fit all their children's luggage inside.

Heather did not think that was a good idea. She did not think that all the luggage could be fit inside the van, and everybody would wind up getting very stressful. The best way was to drive their son to Queen's, come back to Ottawa, and then drive their daughter to University of Toronto in another day.

Since early August, she had been telling George that they should take two separate trips. George definitely did not think that was an efficient way to do it, but he did not reply, as he knew that any objection from him would be futile. He knew Heather well.

When Heather speaks, she speaks the absolute truth. Not only is she right about the past, she is also right about the future. When and if she ever changes her mind, she says it is because the circumstances have changed, and she has to change her plan accordingly. In that sense, she always tells the relative absolute truth.

George cannot forecast the future. He can only choose a path that he thinks would use up the least amount of resource and effort, and has a reasonable probability of success. He cannot predict that a certain idea of his would work out a 100%. However, he will estimate that the path that he follows would have a good chance of being accomplished. In this particular trip, he did check with their children to see how much luggage they each had, and he believed that he could fit everything inside the van.

A couple of days before the trip, Heather brought up the subject again, and said that they should take two trips. George then told her that they would take one trip. If everything did not fit inside the van, then they would leave some of their son's stuff behind, and he would drop off those extra luggage at Kingston the next week. Heather eventually agreed.

The evening before their trip, George asked his son to help him remove the middle seat section of the minivan, and load some of the larger items inside. In the next morning, they loaded the rest of the stuff into the van. George then found out that each of their children had got approximately 50% more luggage than what they originally told him. Fortunately, he had allowed some leeway. In the end, they had to unpack one box, and put the belongings under the rear seat section. Even though the minivan was packed to the rim, George assured that he could still see through the rear window, and everything was safely tied up and would not fall down in case the van had to stop abruptly.

They arrived at Queen's by noon. Much to the credit of Queen's University, the new student orientation was very well organized. In about one hour, they had their son's belongings moved to his room in the student residence. The new students could actually use the phones in their rooms right away. This arrangement was much better than some other universities where a student would have to line up to apply for a phone, and it would then take several days before the phone got hooked up.

The family had lunch at the University's cafeteria. At about 2 pm, they left their son behind at Kingston, and headed for Toronto, arriving at the University of Toronto campus slightly after 4 pm.

George considered the trip very well planned, and everything followed the schedule. For once, Heather agreed.

Example 2 Baby and car seat

James and Cheryl just got married, and they bought a new car. The car had four seats and two doors.

A year later, they had a baby girl. So they had to put a car seat at the back seat of the car. Since the car had only two doors, they found it very inconvenient every time putting the baby in the car seat. They eventually had to sell their two-door car, and bought a four-door car.

James and Cheryl should have foreseen that a baby would be born, and should have bought a four-door car to begin with.

Example 3 Price adjustment

Some stores have a price adjustment policy. If you purchase an item from the store, and then find out later that the item is on sale, you can bring the original sale receipt back, and the difference will be refunded. The period allowed for the price adjustment is usually fourteen days from the date of purchase.

It was December 2007. The father needed a new winter jacket, as the one that he was wearing was getting a bit worn out. The children were always saying that he looked like a homeless in that old jacket. He therefore thought that he should buy a new jacket on or after Boxing Day, when the stores would have sales on. (Boxing Day is December 26, the day after Christmas.) His twenty-year-old daughter offered to go with him, as she believed that she had much better taste than her father. In any case, she is a shopaholic and knows where the best deals are.

Father and daughter went to a shopping centre on January 12, 2008. After looking through a few stores, the daughter picked out a nice jacket for her father. The father tried it on, and it looked good on him. So he bought it at $215, which was 25% off from the regular price.

The daughter, being an observant person, noted that there was a price adjustment policy printed at the back of the receipt. A one-time sale price adjustment is available within fourteen days of the date on the original receipt. She figured that the store would have a pretty good chance of a clearance sale later, and she would keep her eyes out as she went to the shopping centre often.

The following Monday, she told her father that the store had just put out a big sign at the door, saying that it was having a 30% markdown on all items in the store.

A few days later, father and daughter went back to the store, and got a refund of $64.50.

Quite often, it will be very beneficial to estimate the probable value of a path to a destination. For a path that may be rewarding, but requires a lot of effort and does not have a good chance of success, one should consider abandoning it before even starting on the journey. Let us take a look at the following example.

Example 4 Equipment building in graduate schools

For students entering graduate school in the Department of Science or Engineering, it is not uncommon for their supervisors to ask them to build certain equipment for their experiments. Some equipment building may be necessary, as the experiment can be unique and the equipment is not available commercially. Furthermore, it may be a good training to the student, as it may prepare him or her to design and build equipment in the future.

However, sometimes the equipment is commercially available, but the student is asked to build it because the supervisor either does not have the funding to purchase it, or would like to allot the funding for some other purpose. Occasionally, the construction of the equipment may be beyond the capability of the student or even the supervisor, especially when the student is a Master Degree student and is inexperienced. In that sense, it may not be fair for the student to be asked to build the equipment, as he may not be able to get his Degree finished. The following two cases happened at a university in Canada.

Case 1 Building a laser

Ken got his Bachelor Degree in Electrical Engineering at the University. He was happy that he was awarded an NSERC (National Science and Engineering Research Council) postgraduate scholarship to enter graduate school, and he decided to continue with the same university. He found himself a supervisor, who suggested that he should build a laser, and then collect some experimental data, which would form the thesis of his Master Degree.

Ken did not have any experience in building a laser. As a matter of fact, he did not have any experience in building anything. To make it worse, he never got any help from his supervisor, who was one of those professors that sat in his office and did not wander into the laboratory to get his hand dirty. Ken studied the literature about lasers, and ran around talking to people who had experience working with lasers. Unfortunately, after two years, he never got the laser built. Discouraged, he eventually switched to study an MBA (Master of Business Administration).

Case 2 Building a superconducting solenoid

Kwang was funded by the Korean Government to come to Canada to study for a Master Degree in Engineering. He found a supervisor, Professor Lewinsky, who proposed that he should start off by building a superconducting solenoid, and then collect some experimental data for his Master Degree project. A superconducting solenoid is an electromagnetic coil that has negligible electric power consumption at liquid helium temperature (–269° C). It can generate a very stable magnetic field, allowing scientists, and engineers to investigate material properties at very low temperature. To produce the magnetic field, the solenoid needs to be immersed in liquid helium which would be kept in a stainless steel dewar.

As Prof. Lewinsky and Kwang had not got the solenoid working yet, there was no point in purchasing a dewar. But that would not be a problem, as Prof. Lewinsky knew Prof. Martin in the Physics Department. Prof. Martin had a commercial superconducting solenoid in a dewar, and had agreed to loan them the dewar when necessary so that they could test out their solenoid. Prof. Martin had also advised Prof. Lewinsky that private companies had spent years of research in building these solenoids, and he would be better off purchasing a commercial one.

However, Prof. Lewinsky would not listen. He and Kwang eventually got the solenoid built, hauled it over to the Physics Department, took the Physics solenoid out, and put theirs in for testing. Over a period of two years, they tested their solenoid a few times, and never got it to work. By then, the funding from the Korean Government was running out, and Kwang was tired of not making any progress. He quit, and went back to Korea, abandoning any hope of ever getting a Master Degree.

In most Asian countries, when the parents send their kid abroad to study, they expect him or her to come back with a degree. Coming home without a degree is considered to be a disgrace. The engineering professor might not have realized. Not only had he completely changed someone's career, he might have altered someone's life.

Usually, given a path to a solution, we can attempt to increase its probable value. We can increase the probability of success and/or we can increase the final remuneration. For most of the time, it is easier to increase the probability, as the following example will show.

Example 5 Application to medical school

In Canada, medical school is a faculty of a university, and is usually offered as a four-year post-graduate program to students who have received a bachelor's degree.

The competition for applying to medical school is keen. With the aging population, Canada is very much in need of medical doctors. A doctor is guaranteed a good job and high pay. With that anticipation, some undergraduates would devote all their time hitting the book, sparing only a few hours each week for other activities, such as socializing. At a dinner party, one guest asked a medical doctor whether he had a life when he was going through his undergraduate studies before medical school, and was it worth it. The answer was: No, he did not have a life; and Yes, it was all worth it. And it certainly was worth it, as he was making more than a quarter of a million dollars per year, and that was more than four times what a PhD graduate made.

At the universities in North America, a student's quality of performance is represented numerically by his cumulative grade point average (CGPA), which is the weighted mean value of all the grade points that he earned for taking all the courses. The highest CGPA is 4.0, which means that the student gets straight A's in all his courses. For medical schools in Canada, the cutoff CGPA for admission is 3.5. As much more than enough students apply, most students who get interviews actually have CGPA of above 3.6. And one out of two students who are interviewed gets accepted to medical school. Thus, the competition to be admitted to medical school is keen.

Ted very much wanted to be accepted to medical school, both for the money, and for his own interests. He knew that somehow he had to strive to increase his chance. And he had some strategy.

Most universities have pre-requisite courses that the student should take before being admitted to medical school. For example, physics is one of the pre-requisites. But physics courses come in different levels of difficulty. There is a physics course for biology students, and there is one for engineering students, the latter being more difficult. Ted, of course, would take the easier physics course just to fulfil the requirement. He was able to get a GPA of 3.7 for that course. In addition, he was quite careful in choosing all his other courses so that they would not draw down on his total CGPA. In the end, he managed to get a CGPA of 3.9 for all his undergraduate studies. Basically, what he had done was that he worked within the constraints and requirements to maximize his marks.

However, CGPA is not the only criterion that universities are looking at. They also look at the student's score on the Medical College Admission Test (MCAT) as well as his or her extra-curricular involvements.

The MCAT is a computer-based standardized examination designed for prospective medical students in Canada and United States. It is devised to assess problem-solving, written analysis as well as knowledge of scientific concepts. In order to strive for higher score for the MCAT, Ted took an MCAT preparatory course, which consisted of eighty hours of classroom instructions in problem solving on the various subjects tested. Furthermore, he also did practice exam questions on his own for five months. As a result, he was able to get qualifying scores for the MCAT with no problem.

For extra-curricular activities, he volunteered in a hospital for one summer, and worked as a research student for a microbiology professor at the university for another summer. He managed to get good letters of reference from both the hospital, and the professor.

Consequently, he was interviewed by four medical schools, and was accepted by all four. Here, we can see that Ted tried his best to increase his probability of success, and he succeeded.

Example 6 High rise apartment complex

Alice lives in Singapore. She flies to Hong Kong twice a year to visit her father, as well as her brothers and sisters. Quite often they will go out for lunches and dinners. Alice's younger brother, Michael, has a car, and that makes it convenient for them to drive around the city.

On one Saturday, they were driving back to their father's after lunch. When they were quite close to their father's house, they got stuck in a traffic jam, and the cars were inching along. As it happened, there was an entrance/exit of a high rise apartment complex right beside where Michael's car was. (A high rise apartment complex is a group of several high rise apartments built in one residential area.) Alice suggested that Michael drove into the apartment complex, and got out at another entrance/exit of that apartment complex. Michael replied that he had done that before, and there was no other entrance/exit to that complex.

Somewhat to every person's surprise, Alice persisted that Michael should give it a try. Reluctantly, Michael did what she suggested. And, lo and behold, there was indeed another entrance/exit to that apartment complex. So they drove out of that exit, and were able to quickly get back to their father's house.

Alice later explained that she knew something about city planning, and in her estimation, it was highly unlikely that particular apartment complex would have only one entrance/exit.

In this particular example, Alice had induced a general principle from her observations of the high rise apartment complexes in Singapore, and was able to make a deduction in a similar circumstance in a different city. She figured that certain event was unlikely to happen in that specific situation. That evaluation of probability got them to their destination much quicker.

Sometimes, given a path, we may be able to increase the reward, and thereby the probable value, as the following example will show.

Example 7 Lobster buffet

The Fawcetts live in Syracuse in New York State. One summer, the family of four went to Florida for a vacation for two weeks.

Near the hotel where they were staying was a restaurant which served a lobster buffet. Each customer would be served one lobster. After he finished, he could go to the buffet counter, and the server would dish out another lobster for him. And he could go back as many times as he wanted to for more lobsters.

As the whole family loved lobsters, they decided to go and tried out the buffet for dinner. The lobsters turned out to be tender, and not overcooked, and the family enjoyed the meal.

Later on in the evening, the family was still talking about how pleasant the supper was. The father said that he had four lobsters, and he was really full. The twelve-year old son said that he had six. Somewhat surprised, the father asked his son how could he possibly have eaten that much.

The son then explained. Every time the customer went to the buffet counter to get a lobster, the server would serve him a small cup of warm melted butter together with the lobster. The butter would be used to dip the lobster meat into for enhancing the favour. However, butter was also quite filling. So, instead of taking the butter, the son would take a couple of lemon wedges that were provided on a side table, and later squeeze the lemon juice on the lobster to make it taste better. (Lemon juice is quite often used to neutralize the fishy-smelling amines, thus greatly improving the taste of seafood.) The lemon juice could also serve another purpose, as it could stimulate the flow of saliva and gastric juice, making it a good digestive agent. And that was why he could eat six lobsters in one meal.

The father then realized he had just learnt a biology lesson from his son.

In general, in everyday life, if we have several paths to choose, we should estimate the chance of success for each path, and the reward at the end. After calculating the probable value of each path, we should choose a trail that clusters near the one that has the highest probable value, and dropping the ones that lie at the low end.

But then, of course, it does not mean that the path we are taking would definitely lead to an accomplishment. The path is a hypothesis, and it needs experimentation to test its validity. And that is what the scientific method is all about.

Chapter 13

Epilogue

We run into problems every day. Even when we do not encounter any problems, it does not mean that they do not exist. Sometimes, we wish we could be able to recognize them earlier. The scientific method of observation, hypothesis, and experiment can help us recognize, define, and solve our problems.

We should keep our eyes open and our brains alert. And anticipate any problems before they sneak up on us, and catch us off-guard. Not only do we need to recognize the problem, we may have to evaluate and realize how significant they may be. Not detecting the seriousness of a problem can be costly.

Search and collect any relevant information, and come up with several hypotheses as quickly as you possibly can. Choose the one that best explains the present problem situation. (This method of reasoning is called abduction.) Use the hypothesis to forecast what can happen, and then perform experiment to prove that your prediction is indeed correct. Hypothesizing is important, as it will give you a sense of direction. If your hypothesis is incorrect, change your bearing and come up with a new hypothesis. Do spend time and effort to perform the experiments carefully, and verify that your hypothesis is positively unmistaken.

Observation, hypothesis, and experiment do not have to be followed in that order. Proceed in whatever order that is deemed necessary, and repeat as often as is required.

To extrapolate and exploit whatever information we have stored in our brain, we need to visualize the relation among different concepts and integrate them to cope with the problem we face. Creative solutions can arise only if we can see the combination of heretofore-unrelated ideas.

Anyone can come up with bright ideas. It has been shown that creative thinking is no different from ordinary thinking. In any case, whether an idea is trivial or intelligent can be relative. A professional familiar with that particular field may consider the idea simplistic while a layman may regard the concept groundbreaking. We can be considered laymen to many situations in our daily life, and what we consider smart ideas may be viewed as insignificant. However, the important point is to get the problem solved, and not whether the idea is ingenious or not. Occasionally, we may do better than the professionals.

Look at a problem situation from different perspectives, and seek alternative ways to define a problem. Once the problem is defined, we should search for multiple solutions. If possible, allow enough time to come up with various options. Solutions, just like problem definitions, can be viewed from various angles. Inspiration quite often comes after an incubation period. So, take the time to mull over the problem, as well as various plausible solutions.

Understandably, our experience is limited, and the information that we can gather is finite. That is why some basic scientific knowledge would help. Basic scientific theories underlie and explain many phenomena, allowing us to cope with completely new situations that we may have no experience with. In addition, knowing some mathematics, even simple arithmetic, is very beneficial. Some problems simply cannot be solved in a hand-waving manner. They require mathematical evaluation.

Not only do we need to solve problems that occur in the present, we should attempt to anticipate problems that will happen in the future. That is why forecasting is significant. Thus, we should do short term and long term planning. And then act accordingly.

If there are different paths to a destination, evaluate the probable value (= probability of success × reward) of each path, and choose the path which has one of the highest probable values. Work to increase the probability of success and/or reward if at all possible. Every problem has its own constraints — rules and regulations intrinsic in the problem situation, and time, money and efforts that are inherent with the problem-solver. Try to think of various solutions within the constraints and choose the one that will maximize the reward.

Take risks, and attempt something new. If you don't try, you will never find out and you may miss out some opportunities. Do expect making lots of mistakes. A professor once told his new graduate student to "make as many mistakes as quickly as possible".

Learn from your mistakes and failures. Do not cry over split milk. Instead, prepare for your next challenge. Furthermore, if possible, learn from other people's mistakes, and do not repeat their unsuccessful paths.

Discuss and collaborate with others when possible. Two minds are better than one. Other people have knowledge and experience that you do not have. They may point out information that you are completely unaware of. Also, they may come up with ideas that you have never dreamed of.

Not all problems can be solved, just like not all diseases can be cured (at least, not yet). Quite a number of problems have constraints that are beyond your control. However, you will learn that if you familiarize yourself with the scientific method, and keep practicing it, you will find out that you can solve more problems than you could previously have. Occasionally, you may come up with some bright solutions that are very gratifying.

Being able to solve your own problems would make you feel more accomplished, and let you enjoy a better life.

Bibliography

Adams, James L., *Conceptual Blockbusting, A Guide to Better Ideas*, Third Edition, Addison Wesley (1986).

Adams, James L., *The Care and Feeding of Ideas, A Guide to Encouraging Creativity*, Addison Wesley (1988).

Anderson, Margaret J. and Stephenson, Karen F., *Aristotle, Philosopher and Scientist*, Enslow Publishers Inc. (2004).

Anderson, D. R., Sweeney, D. J. and Thomas, A. W., *Statistics for Business and Economics 9e*, Thompson Southwestern (2005).

Bakar, Osman, *The History and Philosophy of Islamic Science*, Islamic Texts Society (1999).

Baker, Samm S., *Your Key to Creative Thinking, How to Get More and Better Ideas*, A Bantam Book (1962).

Bartlett, Sir Frederic, *Thinking, An Experimental and Social Study*, Unwin University Books (1964).

Beveridge, W. I. B., *The Art of Scientific Investigation*, Vintage Books (1957).

Bransford, John D. and Stein, Barry S., *The Ideal Problem Solver*, W. H. Freeman and Company (1984).

Cajal, Santiago Ramon Y, Translated by Neely Swanson and Larry W. Swanson, *Advice for a Young Investigator*, MIT Press (1999).

Carey, Stephen S., *Beginner's Guide to Scientific Method*, Wadsworth (1994).

Carr, Albert, *How to Attract Good Luck*, Wilshire Book Company (1959).

Chung, Deborah D. L., editor, *The Road to Scientific Success, Inspiring Life Stories of Prominent Researchers*, Volume 1, World Scientific (2006).

Csikszentmihalyi, Mihaly, *Creativity*, HarperCollins (1996).

de Bono, Edward, *The Five-Day Course in Thinking*, A Signet Book (1968).

de Bono, Edward, *The Mechanism of Mind*, Pelican Books (1971).

de Bono, Edward, *Children Solve Problems*, Penguin Books (1972).

de Bono, Edward, *The Use of Lateral Thinking*, Penguin Books (1972).

de Bono, Edward, *Po: Beyond Yes and No*, Penguin Books (1973).

de Bono, Edward, *Lateral Thinking, A Textbook of Creativity*, Penguin Books (1980).

Dombroski, Thomas W., *Creative Problem-Solving, The Door to Progress and Change*, iUniverse (2000).

Frank, Robert H. and Parker, Ian C., *Microeconomics and Behavior*, Second Canadian Edition, McGraw-Hill Ryerson (2004).

Feund, Johm E., *Mathematical Statistics*, Fifth Edition, Prentice Hall, Inc. (1992).

Flesch, Rudolf, *The Art of Clear Thinking*, Collier Books (1968).

Gardner, Martin, *Fads & Fallacies, In the Name of Science*, Dover (1952).

Gelb, Michael J., *How to Think like Leonardo da Vinci, Seven Steps to Genius Every Day*, Delacorte Press (1998).

Gelb, Michael J., *Discover Your Genius, How to Think like History's Ten Most Revolutionary Minds*, HarperCollins (2002).

Ghiselin, Brewster, *The Creative Process*, A Mentor Book (1952).

Gordon, William, J. J., *Synectics, The Development of Creative Capacity*, Collier Books (1969).

Higgins, James M., *101 Creative Problem Solving Techniques*, New Management Publishing Company, Inc. (1994).

Hoffmann, Banesh, *Albert Einstein, Creator & Rebel*, Plume (1972).

Jardine, Lisa, *Ingenious Pursuits: Building the Scientific Revolution*, Anchor (2000).

Jones, Morgan D., *The Thinker's Toolkit — Fourteen Powerful Techniques for Problem Solving*, Three Rivers Press (1998).

Killeffer, David H., *How Did You Think of That?: An Introduction to the Scientific Method*, Anchor (1969).

Koestler, Arthur, *The Act of Creation*, Pan Books Ltd (1970).

Kramer, Stephen P., *How to Think Like a Scientist, Answering Questions by the Scientific Method*, Thomas Y. Crowell (1987).

Kuhn, Thomas, *The Structure of Scientific Revolution*, Third Edition, University of Chicago Press (1996).

LeBoeuf, Michael, *Imagineering, How to Profit from Your Creative Powers*, Berkley Books (1986).

Medawar, Peter B., *Advice to a Young Scientist*, Basic Books (1979).

McGee, Harold, *On Food and Cooking, The Science and Lore of the Kitchen*, Simon & Shuster (1984).

McGee, Harold, *The Curious Cook, More Kitchen Science and Lore*, Hungry Minds Inc. (1990).

Michalko, Michael, *Cracking Creativity, The Secrets of Creative Genius*, Ten Speed Press (2001).

Michalko, Michael, *Thinkertoys, A Handbook of Creative-Thinking Techniques*, Second Edition, Ten Speed Press (2006).

Nasar, Sylvia, *A Beautiful Mind, The Life of Mathematical Genius and Nobel Laureate John Nash*, Touchstone (1998).

Nasr, Seyyed Hossein, *Islamic Science: An Illustrated Study*, World of Islam Festival Publishing Company Ltd (1976).

Oech, Roger von, *A Kick in the Seat of the Pants, Using Your Explorer, Artist, Judge, & Warrier to be More Creative*, Harper & Row (1986).

Oech, Roger von, *A Whack on the Side of the Head, How You can be More Creative*, Warner Books (1990).

Ogle, Richard, *Smart World: Breakthrough Creativity and the New Science of Ideas*, Harvard Business School Press (2007).

Osborn, Alex F., *Applied Imagination, Principles and Procedures of Creative Problem-Solving*, Third Revised Edition, Charles Scribner's Sons (1963).

Park, Robert L., *Voodoo Science: The Road from Foolishness to Fraud*, Oxford University Press, Reprint Edition (2001).

Polya, G., *How To Solve It*, Second Edition, Doubleday Anchor Books (1957).

Prince, George M., *The Practice of Creativity, A Manual for Dynamic Group Problem Solving*, Collier Books (1970).

Ridley, Matt, *Francis Crick*, HarperCollins (2006).

Sawyer, R. Keith, *Explaining Creativity, The Science of Human Innovation*, Oxford University Press (2006).

Sawyer, R. Keith, *Group Genius: The Creative Power of Collaboration*, Perseus Books Group (2007).

Siler, Todd, *Think like a Genius*, Bantam Books (1996).

Siu, R. G. H., *The Tao of Science, An Essay on Western Knowledge and Eastern Wisdom*, The MIT Press (1976).

Snyder, Paul, *Toward One Science, The Convergence of Traditions*, St. Martin's Press, Inc. (1978).

Sobel, Dava, *Longitude: The True Story of a Lone Genius Who Solved the Greatest Scientific Problem of His Time*, Walker Publishing Company, Inc. (1995).

Stachel, John, *Einstein from 'B' to 'Z'*, Birkhauser Boston (2001).

Taylor, Edwin F., and Wheeler, John Archibald, *Spacetime Physics*, W. H. Freeman and Company (1966).

Thompson, Charles "Chic", *What a Great Idea, The Key Steps People Take*, HarperPerennial (1992).

Thomson, Sir George, *The Inspiration of Science*, Anchor Books (1968).

Thorpe, Scott, *How to Think Like Einstein: Simple Ways to Break the Rules and Discover Your Hidden Genius*, Sourcebooks, Inc. (2000).

Treffinger, Donald J., Isaksen Scott G. and Dorval Brain K., *Creative Problem Solving, an Introduction*, Third Edition, Prufrock Press Inc. (2000).

Watson, James D., *The Double Helix*, The New American Library, Inc. (1968).

Weisberg, Robert W., *Creativity: Understanding Innovation in Problem Solving, Science, Invention, and the Arts*, John Wiley & Sons, Inc. (2006).

Whitfield, P. R., *Creativity in Industry*, Penguin Books (1975).

Wickelgren, Wayne A., *How to Solve Problems, Elements of a Theory of Problems and Problem Solving*, W. H. Freeman and Company (1974).

Wilson, Jr., E. Bright, *An Introduction to Scientific Research*, Dover (1990).

Woodall, Marian K., *Thinking on Your Feet: How to Communicate under Pressure*, Professional Business Communications (1996).

Youngson, Robert, *Scientific Blunders, A Brief History of How Wrong Scientists can Sometimes be*, Robinson Publishing Ltd (1998).

Index